# Math Everywhere

## ONTARIO EDITION

6

### SENIOR MATH EVERYWHERE AUTHORS
Peter Rasokas, Barry Scully, Jan Scully, Bryan Szumlas

### MATH EVERYWHERE K–8 AUTHOR TEAM

Brendene Barkley  Betty McKendry
Luisa Busato  Linda Miller
Geoff Cainen  Emma Mills Mumford
Tara Cook  Carla Pieterson
Garey Edgar  Pamela Quigg
Donna Green  Maureen Rousseau
Liz Holder  Robert Stoddart
James King  Lori Wiens
Barbara E. Worth

### CONSULTANTS

ASSESSMENT CONSULTANTS Kelly Lantink, Kevin Akins
EDITORIAL CONSULTANT Mary Jean Tyczynski
MATHEMATICS DEVELOPMENT CONSULTANTS Mary Ellen Diamond, Kathleen Nolan
MATHEMATICS LITERACY CONSULTANT Cathy Marks Krpan
TECHNOLOGY CONSULTANTS Marilyn Legault, Doug McKnight
ADVISOR ON ABORIGINAL PERSPECTIVES Ken Ealy

### Elementary Education Advisory Board

CURRICULUM ADVISOR Les Asselstine
LITERACY ADVISOR David Booth
STAFF DEVELOPMENT ADVISOR Rod Peturson

Harcourt Canada

Harcourt Canada

Orlando  Austin  New York  San Diego  Toronto  London

**National Library of Canada Cataloguing in Publication Data**
Main entry under title:
    Math everywhere 6
ISBN 0-7747-1540-5
    1. Mathematics.  I. Rasokas, Peter, 1949–

Editorial Project Manager: Ian Nussbaum
Developmental Editors: Sasha Patton, Brett Savory, Joanne Close, Eliza Marciniak,
    Todd Mercer, Erynn Prousky, Elizabeth Salomons
Production Editor: Tricia Carmichael
Production Coordinator: Cheri Westra
Production Assistant: Agnieszka Mlynarz
Permissions Editor and Photo Research: Karen Becker
Art Direction and Design: Sonya V. Thursby/Opus House Incorporated
Composition: Susan Purtell
Printing and Binding: Transcontinental Printing Inc.
Cover Image: © Larry Prosor/Superstock

∞ Printed in Canada on acid-free paper.

1 2 3 4 5    07 06 05 04 03

Welcome to *Math Everywhere!*

Did you know math is all around you? Math is in your classroom, it's in your home, and it's in your city or town.

Math has also been used throughout history and can be used to plan the future.

At the beginning of this year, in **Start-Up Math**, you will review what you already know about math and prepare for the school year.

In **Unit 1, Math in Space**, you will learn about numbers and patterns as you learn about space exploration.

In **Unit 2, Math and Flight**, you will learn about shapes and measurement, and about numbers and patterns as you learn about fliers, airports and airplanes, and the history of flight.

In **Unit 3, The Math of Starting a Business**, you will learn about numbers, data, patterns, and measurement as they relate to starting a business.

At the end of the year, in **Celebrating Math**, use what you have learned about math this year to plan a class picnic.

At the end of each chapter, you will review what you have learned and show your understanding of math by doing a wrap-up activity.

You will play games, solve riddles, and solve problems while learning all about math!

You will see yellow shapes in some parts of the book:
- tells you that you will be solving a problem.
- lets you know that your answers will show what you understand about math.
- shows questions that will let you apply what you know about math.
- tells you that you will communicate what you know about math.

We hope you enjoy using this book and that you have fun learning that math **is** everywhere.

# Contents

## Start-Up Math

Start-Up Math introduces you to *Math Everywhere* and lets you show what you already know about math.

## Unit 1

### Math In Space

**Chapter 1**
**Math in Earth's Orbit**                               21

This chapter focuses on *Number Sense and Numeration*. This chapter also touches upon *Patterning and Algebra* and *Data Management and Probability*.

## Chapter 3
## Math Beyond Mars                                  106

This chapter focuses on *Number Sense and Numeration*. This chapter also touches upon *Patterning and Algebra*.

## Chapter 6
## Math and Flight: Past and Present     249

This chapter focuses on *Geometry and Spatial Sense*.
This chapter also touches upon *Number Sense
and Numeration*, *Patterning and Algebra*,
and *Measurement*.

## Unit 3

# The Math of Starting a Business

This chapter focuses on *Number Sense and Numeration*. This chapter also touches upon *Measurement* and *Geometry and Spatial Sense*.

## Celebrating Math

Celebrating Math lets you show what you have learned about math while celebrating the end of the year.

# Start-Up Math

Welcome to *Math Everywhere*. To begin the year, you will use what you already know to complete several mathematics activities. The Start-Up activities are designed to check your understanding of math concepts, and prepare you for the work you will be doing in *Math Everywhere 6*.

Answer the following questions to start you thinking about math:

1. If the answer is 1, what is the question? Write some addition questions that each have the answer "1."

2. Make a pattern by writing a series of questions so that the answer to each question is 10 more than the one before it.

3. If the answer is 100, what is the question? Write several questions that each involve two operations.

## Lesson 1

# Thinking About Math

## Get Started

"Get Started" gives you a few quick activities to prepare you for the math lesson.

Take a moment to think about the title of this book, *Math Everywhere 6*. Share with a classmate what you think this title means.

## Build Your Understanding

### Math Web

"Build Your Understanding" gives you a clue about what you will be learning or doing in the lesson.

You Will Need
• sheet of legal-size paper
• pens or markers

Work individually.

1. Draw an oval shape in the middle of the sheet of paper. Inside the oval, write the word "math."

2. Draw five branches shooting out from the oval. Label the branches "numbers," "measurement," "geometry," "patterns," and "data management."

3. Now link as many of your own ideas as you can to the branches on your math web. Use pens or markers to decorate your web.

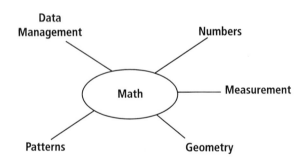

## Tip

If you are stuck for thoughts, look through *Math Everywhere 6* for ideas.

## What Did You Learn?

The "What Did You Learn?" sections are designed to help you think about what you have learned in the lesson.

1. Share your math web with a classmate. Then post it on a bulletin board. Remember to include your name.

2. In your math journal, record things you find challenging in math and things you find easy in math.

## Practice

The "Practice" sections include many different types of activities that you can do to practise or review what you have learned, or to challenge you further.

Make an acrostic poem for the title of your math textbook. Find a math word starting with each of the letters in "Math Everywhere."

M ultiplication
A
T
H

E
V
E
R
Y
W
H
E
R
E

**Vocabulary**

**acrostic poem:** A poem in which the first letters of each line form a new word or phrase

Have a class debate. Divide the class into two groups. One side will argue that math is everywhere, and the other side will argue that math is not everywhere.

## Lesson 2

# Finding Missing Numbers

## Get Started

Work with a partner to find the missing numbers.

1. $7 \times \blacksquare = 49$

2. $\blacksquare \times 9 = 81$

3. $234 + \blacksquare = 500$

4. $4563 - 3452 = \blacksquare$

5. $10 \times \blacksquare = 170$

6. $\blacksquare \times \blacksquare = 18\,000$

7. $42 \div 6 = \blacksquare$

8. $1234 \div \blacksquare = 617$

9. $267 - \blacksquare = 214$

10. $112 \div \blacksquare = 56$

11. $\blacksquare + 588 = 1157$

12. $\blacksquare \div 17 = 5$

13. $5 \times \blacksquare = 250$

14. $88 \div \blacksquare = 11$

15. $\blacksquare \times 9 = 99$

16. $45 \times 8 = \blacksquare$

17. $382 \div 2 = \blacksquare$

18. $56 \div \blacksquare = 8$

## Build Your Understanding

### Build Numbers

You Will Need
• ruler
• pens or markers

Work individually.

1. Use your ruler and pens or markers to draw a ladder that is 20 cm tall.

2. Space the steps on the ladder 2 cm apart.

3. Think of a number between 100 and 1000. Write it down.

4. Now think of a math sentence for each step of the ladder so that all the answers equal the number you chose.

5. Repeat this activity using a different number.

## What Did You Learn?

1. Share your math ladder with a classmate.

2. Which operation was easiest to use when making questions for your ladder? Why?

3. Which operation was more challenging to use when making questions for your ladder? Why?

4. What will you do differently next time you make a math ladder?

5. How many steps will there be on a 30 cm ladder if the steps are 2 cm apart? How can you use your ruler to help you find the answer?

## Practice

1. $2345 + \blacksquare + 4531 = \blacksquare$

2. $4532 - \blacksquare = 3421$

3. $\blacksquare \times 17 = 51$

4. $\blacksquare \times \blacksquare = 121$

5. $1624 \div 4 = \blacksquare$

6. $562 + \blacksquare + 89 = 696$

7. $8956 - \blacksquare + 8900 = \blacksquare$

8. $62 \times \blacksquare = 310$

Write a multiplication sentence and a division sentence for each answer below. Challenge yourself by not using the number 1 in your math sentences.

9. 32     10. 64     11. 144     12. 500     13. 10 000

### Extension

14. Show these numbers on a place-value chart. The first one has been done for you.

34 576

| Ten Thousands | Thousands | Hundreds | Tens | Ones |
|---|---|---|---|---|
| 3 | 4 | 5 | 7 | 6 |

a) 2305     b) 46 007     c) 20 762     d) 9040     e) 100 000

**Lesson 3**

# Exploring Three-Dimensional Shapes

## Get Started

Name these three-dimensional shapes.

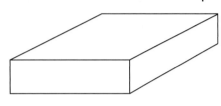

A

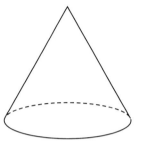

B

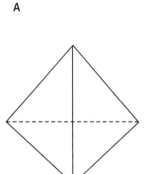

C

D

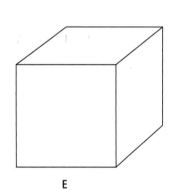

E

## Build Your Understanding

### Tower Tally

You Will Need
• three-dimensional solids

Work in a small group.

1. Assign a number between 1000 and 900 000 to each three-dimensional solid. Keep a record of your work.

2. Have each person in the group take turns making a tower using the solids.

3. Draw a picture of each tower and calculate its value.

4. Make math sentences for each picture.

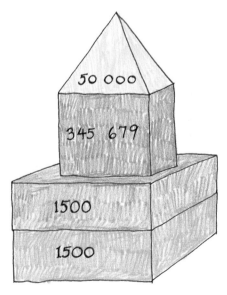

1500 + 1500 + 345 679 + 50 000 = ?

## What Did You Learn?

1. Why is a rectangular prism called a rectangular prism?

2. How are all pyramids the same? How are they different?

3. Which tower had the greatest value?

4. Which tower had the least value?

5. What is the total value of all the towers your group made?

6. Order the values of the towers from greatest to least.

## Practice

Solve these questions in your notebook:

1. 1234 + 889 345 = ▨

2. 360 310 + 893 = ▨

3. 999 001 − 999 = ▨

4. 65 324 − 5604 = ▨

5. 9902 − 8934 = ▨

6. 124 598 + 345 363 = ▨

7. Change each addition question to a subtraction question. Change each subtraction question to an addition question.

8. Write the steps you followed to change each question.

### Extension

You Will Need
• three-dimensional solids

9. Draw a net of a cube. Colour the top and bottom faces of your cube red. Cut out your net and fold to check whether you coloured the correct faces.

10. Count the edges, faces, and vertices of a three-dimensional solid of your choice. How many of each did you find? Exchange solids with a classmate. Did you get the same totals?

## Lesson 4
# Exploring Measurement

## Get Started

**1.** How long is each item in millimetres?

**a)**

mm ruler

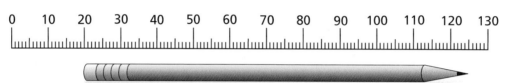

**b)**

mm ruler

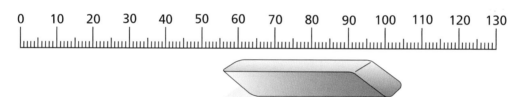

**c)**

mm ruler

**d)**

mm ruler

**2.** How long is each item in centimetres?

## Build Your Understanding

### Measure With Standard Units

You Will Need
- coloured chalk
- metre sticks
- broom

Work with a partner.

1. Sweep a 3 to 5 m stretch of sidewalk where you will be working.

2. Working together, use your chalk to draw a city scene with different-sized buildings.

3. Use two different standard units to measure the height of each building.

4. In your notebook, make a chart to record your findings.

## What Did You Learn?

1. Arrange the objects from "Get Started" in order from longest to shortest.

2. Name five different linear measurements, and give an example of when you would use each unit of measurement.

3. Using numbers, pictures, and words, show the relationship between millimetres, centimetres, decimetres, and metres.

4. Would there be more centimetres or decimetres in the height of a stop sign?

**1.** Draw the following:

　**a)** a square with a perimeter of 16.8 cm

　**b)** a triangle with 1 dm sides

　**c)** a quadrilateral that has a perimeter of 21 cm, and each side is a different length

　**d)** a trapezoid that has 2 sets of parallel lines. The total length of all of the sides is 24 cm.

**2.** Measure the circumference of each circle in the target below.

**Tip**

You can use a piece of string to measure the circumference of a circle. Wrap a piece of string around the circle, then lay the string flat on a ruler to measure the length.

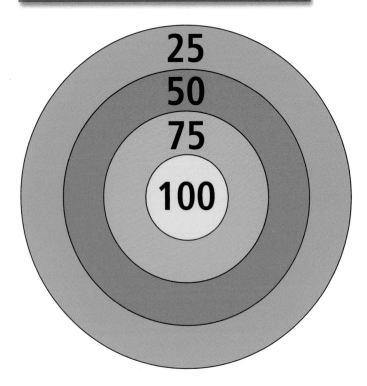

25

50

75

100

# Problems to Solve

Here are some interesting problems for you to solve. For each problem, a helpful problem-solving strategy is given. Later in the year, when you have learned more strategies, you will have the chance to choose the strategies you use.

**Problem 1**

## How Does It Measure Up?

STRATEGY: ACT IT OUT

Playing a role or acting out a problem lets you see the problem more clearly and helps you figure out a solution.

OBJECTIVE:

Measure using a nonstandard unit of measurement

### Problem-Solving Steps

There are four steps you can follow to help you solve a math problem. You will be reminded of these steps throughout the year:

1. **Understand the problem:** Rewrite the problem in your own words. If you can, draw a picture of the problem. List or highlight important numbers or words.

2. **Pick a strategy:** For example, "Act It Out," "Draw a Picture," "Use Objects," and "Guess and Check."

3. **Solve the problem:** Use a strategy to solve the problem. Describe all steps using math words and/or symbols. Try a different strategy if you need to. Organize the results using a diagram, model, chart, table, or graph.

4. **Share and reflect:** Did the strategy you picked work? Would a different strategy also work? Does your solution make sense? Could there be more than one answer to the problem? How did other people in your class solve the problem?

## Problem

The caretaker in your school has lost her tape measure. She needs your help to find out the rough dimensions of your classroom. Work in a group to figure out the length and width of your classroom. You may not use a ruler, metre stick, or tape measure.

## Reflection

1. What did you know about the problem before you started to solve it?

2. What did you need to figure out?

3. What nonstandard unit of measurement did you use?

4. Do you think the "Act It Out" strategy was a good strategy to use? Why?

5. How will you record a solution? Share your solution with your teacher.

## Extension

You Will Need
• metre stick

Measure the length and width of the classroom in centimetres and metres. How do these dimensions compare to the measurements you made before?

**Problem 2**

# Area and Perimeter of Sculptures

STRATEGY: DRAW A PICTURE OR DIAGRAM

You can draw simple pictures or diagrams to help you solve a problem. They let you see how the parts of a problem work together.

OBJECTIVE:

Demonstrate an understanding of area and perimeter

## Problem

You Will Need
- 1-cm grid paper

An artist has just finished 4 sculptures and she wants to exhibit them at an art gallery. The gallery needs to know how much space will be needed for her work. The table shows the measurements for the bases of the sculptures.

| Sculpture | Measurement |
|-----------|-------------|
| 1 | perimeter = 24 cm |
| 2 | area = 32 cm$^2$ |
| 3 | perimeter = 48 cm |
| 4 | area = 24 cm$^2$ |

1. Using grid paper, find the area for sculpture 1 and for sculpture 3.

2. Using grid paper, find the perimeter for sculpture 2 and for sculpture 4.

base

## Reflection

**1.** What did you know about the problem before you began to solve it?

**2.** What did you need to figure out?

**3.** Do you think the "Draw a Picture or Diagram" strategy was a good strategy to use? Why?

**4.** Is there more than one answer? Explain.

**5.** Compare your pictures or diagrams with a classmate's. How are they the same, and how are they different?

## Extension

The artist wants to paint her sculpture using 2 different colours. She can choose from 5 colours: red, brown, blue, green, and yellow. What are the chances of her picking red and brown?

**Problem 3**

# Organizing the Olympics

STRATEGY: USE OBJECTS

Using objects can help you organize information so you can see the solution. You can use simple objects, such as blocks or pieces of paper.

OBJECTIVE:

Estimate to make a schedule

## Problem

You Will Need
• counters or different-coloured blocks

Imagine that you are planning an Olympic Games day at school. The events include javelin, shot put, discus, long-distance running, and the 50 m sprint. Each category is divided into separate events for girls and boys.

### Tip

Use counters or different-coloured blocks to help you organize the events.

1. Show the different ways that you can schedule these 5 events throughout 1 school day. Figure out as many possibilities as you can.

2. How can these events be organized so that each running event is never right before or right after another running event? Show as many possibilities as you can think of.

## Reflection

1. What did you know about the problem when you set out to solve it?

2. What did you need to figure out?

3. Do you think the "Use Objects" strategy was a good strategy to use? Why?

## Extension

1. Your friend threw the javelin 3 times farther than the shot put, which he threw $\frac{1}{2}$ as far as the discus. He threw the discus somewhere between 7.5 m and 12.4 m. Imagine that you measured to find the exact result. How far did he throw the javelin?

2. In a small group, design a problem that could be solved using the "Use Objects" strategy. Record your problem and give it to another group to solve.

## Problem 4

# Marble Probability

STRATEGY: GUESS AND CHECK

One way to solve a difficult problem is to make guesses and then check to see if your answers are correct. You can use this strategy when you are working with large numbers or if the problem involves many pieces of information. Sometimes what you find out as you use the "Guess and Check" strategy leads you to other strategies.

OBJECTIVE:

Estimate the probability of an event

## Problem

**You Will Need**
• marbles or base-ten blocks
• paper bag

You and your classmates will play the marble grab game. Each player will get a turn to reach into the bag and pull out a marble. There are 7 marbles in the bag: 2 yellow, 3 red, and 2 green. Each time a player reaches into the bag, he or she must first predict the colour of marble that he or she will pull out. The winner is the player who predicts correctly most often in 10 turns. After each pick, the marble is replaced so there are always 7 marbles in the bag. Use a tally sheet like this one to record each player's results.

| Predicted colour | Actual colour | Was your prediction correct? (Yes or No) |
|---|---|---|
| | | |
| | | |

## Reflection

**1.** How did you come up with your guesses?

**2.** How close were your guesses to the actual results?

**3.** Do you think the "Guess and Check" strategy was a good strategy to use? Why?

## Extension

**1.** Imagine there are 867 unit cubes in the bag. 256 are red. The rest are yellow or green. How many yellow cubes and green cubes might there be? Provide one possible solution.

**2.** Share your answer with a classmate.

## Problem 5

# Summer Vacation Statistics

STRATEGY: WORK BACKWARDS

Sometimes the best way to solve a problem is to begin with the answer or information at the end of the problem and work backwards toward the beginning.

OBJECTIVE:

Demonstrate an understanding of percentage and fractions

## Problem

In math class, your teacher presents you with the following question: "What fraction of the students in the class stayed home during summer vacation and what fraction went away to camp?" There are 11 boys in your class, and 8 went to camp. There are 13 girls, and 7 went to camp. Find the answer to the teacher's question.

## Reflection

1. What information did you know before you began to solve the problem?

2. What did you need to figure out?

3. Do you think the "Work Backwards" strategy was a good strategy to use? Why?

4. Explain your solution to a classmate. Do you have different answers? Why?

You are carrying a box of 50 muffins into the school for a fundraising event. Just as you are going up the front steps, you trip and the box goes flying through the air. $\frac{1}{2}$ of the muffins remain in the box, $\frac{1}{4}$ fall onto the stairs, $\frac{1}{8}$ land in the bushes beside the stairs, and 2 muffins land on the path leading to the stairs. Some of the muffins are broken into pieces during the accident.

1. What is the number of muffins found in each location?

2. What is the number of muffins that were still left in the box? What percentage of the total is this?

3. If the remaining muffins then fell into the 4 locations equally, what might the new total numbers be?

# Unit 1
# Math in Space

The Grade 6 math class is on its way to the space exhibit at the Science Centre. You will find out how math is used in space exploration.

In Chapter 1, you will look at math in Earth's orbit. You will practise operations with large numbers, whole numbers, decimals, and money amounts. You will estimate and solve equations, and learn about multiples and factors. At the end of the chapter, you will design and plan your own space mission!

In Chapter 2, you will see how math helps us understand both our moon and Mars. You will study patterns, multiples, factors, prime numbers, missing terms, decimals, and percents. At the end of the chapter, you will design your own game about the solar system.

In Chapter 3, you will look at math beyond Mars and space beyond our solar system. You will practise solving equations, and study large numbers, missing values, data collection and interpretation, and mean, median, and mode. At the end of the chapter, you will prepare a project about a planet or star of your choice, using all of the math and information you have learned.

# Math in Earth's Orbit

Math is a big part of travelling through space. You will learn some of the math skills that astronauts use in space.

In this chapter, you will

- add and subtract large numbers
- multiply and divide whole numbers and decimals
- learn the order of operations
- estimate answers to problems, and find mean and mode
- add and multiply money amounts
- learn about multiples and factors

At the end of this chapter, you will use all of the math skills you have learned to plan your own space mission!

Answer these questions to prepare for your adventure:

1. What is a rule you use when adding and subtracting decimal numbers? What is a rule you use when multiplying and dividing decimal numbers?

2. What is a multiple? Can you describe what a factor is? Explain your answers using numbers, pictures, and words.

# Adding and Subtracting Four-Digit Numbers

*SPACE LOG*

*PLAN:*
*You will use addition and subtraction to compare the sizes of Earth, the moon, and the sun.*

*DESCRIPTION:*
*The universe is a very big place. Even the biggest objects we see in the sky are very, very small compared to the rest of the universe. Math helps us to measure the universe and all its parts.*

## Get Started

The biggest objects in our sky are the sun and the moon. When you look at them from Earth with just your eyes (that is, without using a telescope), they look about the same size. But the sun is actually many, many times bigger than the moon.

| Object in Sky | Diameter |
| --- | --- |
| Sun | 1 390 000 km |
| Earth | 12 756 km |
| Venus | 12 100 km |
| Moon | 3 476 km |
| Mercury | 4 880 km |
| Mars | 6 794 km |
| Pluto | 2 300 km |

## Vocabulary

**diameter:** A line segment joining two points on the circumference of a circle and passing through the centre of the circle

The moon, which looks like one of the biggest objects in our sky, is actually much smaller than the planet Venus. Venus looks about the same size as a star. This is because Venus is much farther from Earth than the moon is. The sun is 400 times larger than the moon, but looks about the same size because it is 400 times farther away.

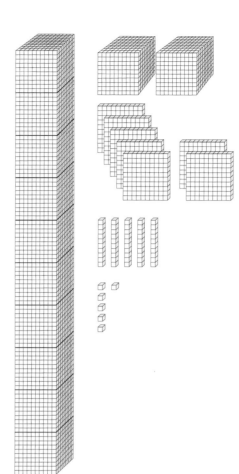

You Will Need
• base-ten blocks

1. Use base-ten blocks to represent the diameters of the moon and the five smallest planets.

2. Copy and complete this place-value chart in your notebook. The first row has been done for you.

| Diameter | (km) | Ten Thousands (blocks) | Thousands (blocks) | Hundreds (flats) | Tens (rods) | Ones (cubes) |
|---|---|---|---|---|---|---|
| Earth | 12 756 | 1 | 2 | 7 | 5 | 6 |
| Moon | 3 476 | | | | | |
| Venus | 12 100 | | | | | |
| Mercury | 4 880 | | | | | |
| Mars | 6 794 | | | | | |
| Pluto | 2 300 | | | | | |

**Technology**

Use a spreadsheet computer program to create a numeric crossword puzzle using four- and five-digit addition and subtraction questions.

## Add and Subtract Data

1. According to the chart, which planet is the largest? Which is the smallest?

2. How much smaller is the moon than Earth? How did you solve this problem?

3. How much bigger is Mars than Mercury?

4. Which is bigger: Pluto or the moon? By how much is one bigger than the other?

### Journal

Which operation do you find more helpful to solve the problems, addition or subtraction? Why?

### Booklink

**Is a Blue Whale the Biggest Thing There Is?**
By Robert E. Wells (A. Whitman: Morton Grove, IL, 1993). This book compares the measurements of big things (for example, a blue whale), bigger things (for example, a mountain), and biggest things (for example, the universe).

## What Did You Learn?

1. Explain how you add three numbers, using numbers, pictures, and words.

2. Explain how you subtract a four-digit number from another four-digit number, using numbers, pictures, and words.

3. Does breaking a number down in a place-value chart help you to understand how big the number is? Explain your answer.

## Practice

Add the following:

| | | | | |
|---|---|---|---|---|
| **1.** 6802 <br> + 2343 | **2.** 7204 <br> + 2244 | **3.** 9321 <br> + 2319 | **4.** 1578 <br> + 4523 | **5.** 8324 <br> + 4589 |
| **6.** 4823 <br> 3602 <br> + 1456 | **7.** 1267 <br> 4420 <br> + 7902 | **8.** 5489 <br> 2301 <br> + 1121 | **9.** 7482 <br> 5460 <br> + 3121 | **10.** 9080 <br> 6323 <br> + 5560 |

Subtract the following:

**11.**  5400
  – 3101

**12.**  6444
  – 3120

**13.**  2231
  – 1004

**14.**  4650
  – 2245

**15.**  8324
  – 3112

**16.**  5230
  – 2397

**17.**  9823
  – 3041

**18.**  9133
  – 1003

**19.**  8030
  – 6499

**20.**  3547
  – 2987

## Extension

**21.** How many thousand blocks would you need to represent the diameter of the sun? Would there be enough room in your classroom for all of them? Explain.

**22.** Which is larger, Venus or Earth? By how much is one planet larger than the other?

Venus

Earth

## A Math Game to Play

You Will Need
• number cube

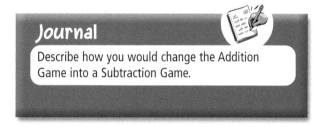

**Journal**

Describe how you would change the Addition Game into a Subtraction Game.

### How to Play

Use a number cube to create addition questions.

**1.** Roll the number cube four times. Write down each number to create a four-digit number.

**2.** Then roll the number cube four more times and write down those numbers. You now have two four-digit numbers.

**3.** Use these two numbers to create and solve an addition question.

**4.** Take turns with a classmate creating and solving questions.

**Lesson 2**

# Solving Multiplication and Division Problems

SPACE LOG

PLAN:
You will add, subtract, multiply, and divide your way through the space exhibit at the Ontario Science Centre.

DESCRIPTION:
Scientists help us understand the universe. Science centres, like the Ontario Science Centre, teach us much of what scientists have learned. The Ontario Science Centre has many different exhibits, including a space exhibit. Among the many things to see and do in the space exhibit at the Ontario Science Centre is a rocket chair you can use to explore the moon, or you can take part in a journey to Mars!

## Get Started

Answer these questions using numbers, pictures, and words.

1. The class is taking a trip to see the space exhibit at the Science Centre. There are 26 students, 1 teacher, and 2 parents going on the trip. The bus fare is $2.00 each way. The Science Centre admission fee is $10.00 for each student, and $15.00 for each adult. How much will the trip cost?

2. Imagine that there are 150 displays in the space exhibit at the Science Centre. Each display takes about 5 min to view. The class has 3 h to spend at the exhibit. How many displays can they see as a group?

3. If the students pair up and separate from the group, and each pair spends 9 min viewing each display, how many displays will be viewed by each pair in 3 h?

**4.** There is a presentation in the space exhibit that is 45 min long. If half of the class goes to the presentation as a group, and the rest of the class as a group continues to look at the displays at the rate of 5 min per display, what is the total number of displays that will be viewed by the time the presentation is finished?

**Journal**

Describe when you used multiplication and when you used division to solve these problems. For which problems could you use either operation?

When you divide a number by a two-digit number, you can use multiplication to help you. When you don't know how many times a two-digit number can go into another number, think of a multiplication question that you do know.

$$16\overline{)48}$$
$$-32 \quad | \quad \textbf{x 2}$$
$$16$$
$$-16 \quad | \quad \textbf{x 1}$$
$$0$$

Now we add the numbers in the right column.

2 + 1 = 3

$$16\overline{)48}$$
$$-32 \quad | \quad \textbf{x 2}$$
$$16$$

In this example, we might not know how many times 16 goes into 48, but we do know that 16 x 2 = 32.

Then we subtract 32 from 48.

$$16\overline{)48}$$
$$-32$$
$$16$$
$$-16$$
$$0$$

with **3** written above.

Therefore, the answer is 3.

$$16\overline{)48}$$
$$-32 \quad | \quad \textbf{x 2}$$
$$16$$
$$-16 \quad | \quad \textbf{x 1}$$
$$0$$

We know that 16 goes into 16 once. Then we subtract 16 from 16.

## Multiply and Divide

Ajay, Mahalia, and Eric visited an interesting display.

The display was a scale model of part of our solar system. Earth was about the size of a pea, and the moon was about the size of a grain of rice. They were about 10 cm apart. The sun stood about 150 m away at the end of the room, and was about as big as a basketball. There wasn't room inside the museum to represent the other planets to scale.

**Solar System**

Pluto
Neptune
Uranus
Saturn
Jupiter
Mars
Moon    Earth
Sun    Venus
Mercury

If you could fly to the sun on a 747 jet, at a speed of 1000 km/h, it would take you 18 years to get there.

1. The space shuttle averages a speed of 27 000 km/h. If it takes 18 years for a 747 jet to fly to the sun, how long would it take the space shuttle?

2. Share with a partner how you solved this problem. Is there more than one way to find the solution?

3. Work with the same partner to write more problems. Exchange problems with another pair of students and solve them.

## What Did You Learn?

**1.** How did you use multiplication to solve the problems?

**2.** How did you use division?

**3.** How did the problems you and your partner created compare with another pair's problems?

**4.** Did you find the other pair's problems easy or challenging to solve? Explain your answer.

## Practice

Multiply the following:

| **1.** | **2.** | **3.** | **4.** | **5.** |
|---|---|---|---|---|
| 68 | 74 | 91 | 15 | 84 |
| x 43 | x 24 | x 19 | x 45 | x 89 |

Divide the following:

**6.** 64 ÷ 16 = ▓    **7.** 168 ÷ 24 = ▓    **8.** 48 ÷ 12 = ▓

**9.** 23)‾184‾    **10.** 61)‾549‾

**11.** If 240 is the answer to a multiplication question, what is a possible question?

**12.** If 120 is the answer to a division question, what is a possible question?

### Extension

**13.** Earth's diameter is 12 756 km and the moon's is 3476 km. About how many moon diameter lengths would fit into Earth's diameter length? Estimate what the solution would be, then figure it out.

**14.** Imagine that the moon is 50 km from Earth, the sun is 50 times farther from Earth than the moon, and a spacecraft travels to the moon at a speed of 10 km/h. If a new spacecraft travelled to the sun 10 times faster than to the moon, which would the astronauts get to first, the sun or the moon?

## A Math Game to Play

You Will Need
• number cube

### How to Play

Use a number cube to create multiplication problems.

**1.** Roll the number cube twice. Write down each number to create a two-digit number.

**2.** Then roll the number cube twice more and write down those numbers. You now have 2 two-digit numbers.

**3.** Use these numbers to create and solve a multiplication question.

**4.** Take turns with a classmate creating and solving questions.

### Booklink

**Moira's Birthday** by Robert N. Munsch and Michael Martchenko (Annick Press: Toronto, ON, 1987). Moira is planning her birthday party, and she invites more and more children until she has invited all the children in her whole school! You can solve different problems like "If Moira gave 29 of her 94 presents back to the children who came early to help with the decorations, how many presents did she have left?"

### Technology

Use a desktop publishing program to create some signs to post in your classroom and around the school outlining tips to help other students with problem solving.

## Lesson 3

# Adding and Subtracting Decimals

SPACE LOG

PLAN:
You will add and subtract decimals to find out about Earth's oceans.

DESCRIPTION:
Of all the planets in our solar system, scientists know the most about Earth. One of the unique things about our planet is that water covers a large part of its surface.

## Get Started

Earth's surface area is $\frac{7}{10}$ water.

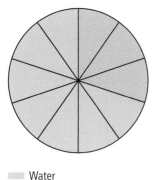

Water
Land

$\frac{7}{10}$ is the same as $\frac{70}{100}$ and $\frac{700}{1000}$.

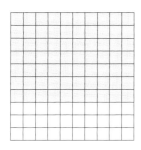

$\frac{7}{10} =$

$\frac{7}{10} =$

| Ones | Tenths |
|------|--------|
| 0 . | 7 |

### Tip
For $\frac{700}{1000}$, imagine that each square in the hundreds grid to the left represents 10.

**1.** Copy and complete the place-value chart below for $\frac{7}{10}$:

| Ones | Tenths | Hundredths | Thousandths |
|------|--------|------------|-------------|
| . |  |  |  |

**2.** Show the following fractions on your place-value chart:

**a)** $\frac{20}{100}$   **b)** $\frac{4}{10}$   **c)** $\frac{56}{100}$

**d)** $\frac{8}{10}$   **e)** $\frac{4}{10}$   **f)** $\frac{3}{1000}$

**g)** $\frac{12}{1000}$   **h)** $\frac{345}{1000}$   **i)** $\frac{629}{1000}$

### Vocabulary

**decimal:** A number based on 10 with one or more digits to the right of the decimal point; for example, 0.36

**denominator:** The bottom part of a fraction that tells how many parts are in the whole; for example, $\frac{3}{4}$

**fraction:** A number that names a part of a whole

**numerator:** The top part of a fraction that tells how many parts are being referred to; for example, $\frac{3}{4}$

## Build Your Understanding

## Display and Compare Fractions

Earth has many bodies of water. Four of the largest are the Pacific Ocean, the Atlantic Ocean, the Indian Ocean, and the Arctic Ocean. The following fractions show what proportion of Earth's water each ocean represents. For example, the Pacific Ocean contains $\frac{52}{100}$ of Earth's water.

Arctic $= \frac{1}{100}$   Atlantic $= \frac{24}{100}$   Indian $= \frac{21}{100}$   Pacific $= \frac{52}{100}$

**1.** Put the oceans in order from smallest to largest.

**2.** Match each grid below to the ocean it represents.

A 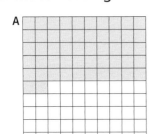   B 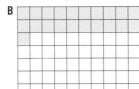   C 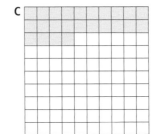   D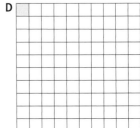

**3.** Match each decimal number to its fraction.

    **a)** 0.52    **b)** 0.21    **c)** 0.24    **d)** 0.01

**4.** Convert each fraction out of 100 to a fraction out of 1000. Now change each fraction out of 1000 to a decimal. What do you notice?

**5.** Copy and complete a place-value chart like the one in Get Started for each number.

## Add Decimals

**1.** Do the decimals add up to 1? Estimate your answer and then copy and complete the chart below in your notebook.

| Pacific | 0.52 |
|---------|------|
| Atlantic | + 0.24 |
| Arctic | + 0.01 |
| Indian | + 0.21 |
| Total | = |

**2.** To solve the problem, add the first two numbers: 0.52 + 0.24. Add your answer to the third number: 0.01. Add your new answer to the last number: 0.21.

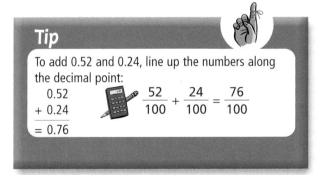

**Tip**

To add 0.52 and 0.24, line up the numbers along the decimal point:

  0.52
 + 0.24
 = 0.76

$$\frac{52}{100} + \frac{24}{100} = \frac{76}{100}$$

**3.** Check your results with a classmate. What did you find out?

**Tip**

The total amount of all the water on Earth would add up to 1. But this answer isn't 1. This is because there are many, many more bodies of water on the planet, all of which make up approximately $\frac{2}{100}$ of the world's water.

## Subtract Decimals

If 0.70 or $\frac{70}{100}$ of Earth's surface area is water, then $\frac{30}{100}$ is land.

1. Add $\frac{70}{100} + \frac{30}{100}$

    $0.70 + 0.30$

2. Another way to determine how much of Earth's surface area is land would be to subtract 0.70 from 1.

3. To subtract 0.70 from 1, line up the decimal points:

$$
\begin{array}{r}
1.00 \\
- \ 0.70 \\
\hline
= 0.30
\end{array}
$$

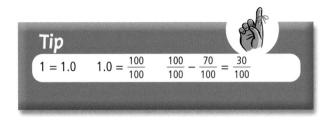

**Tip**

$1 = 1.0 \qquad 1.0 = \frac{100}{100} \qquad \frac{100}{100} - \frac{70}{100} = \frac{30}{100}$

**Journal**

Write down any strategies you use to help you add and subtract decimals.

## Write Other Fractions as Decimals

To write any fraction as a decimal, multiply the denominator by a number to create a new denominator that is 10, 100, or 1000. You must multiply the numerator by the same number.

For example, in the fraction $\frac{4}{5}$, what can we multiply 5 by to get a product of 10?

(5 x 2 = 10)

$\frac{4 \times 2}{5 \times 2} = \frac{8}{10}$

$= 0.8$

In the fraction $\frac{1}{20}$, what can we multiply 20 by to get a product of 100?

$\frac{1 \times 5}{20 \times 5} = \frac{5}{100}$

$= 0.05$

In the fraction $\frac{1}{8}$, what can we multiply 8 by to get a product of 1000?

$\frac{1 \times 125}{8 \times 125} = \frac{125}{1000}$

$= 0.125$

## What Did You Learn?

1. If the area of Earth's oceans were 100 million m² all together, how many million m² would the Atlantic Ocean be?

2. What is the relationship between $\frac{1}{10}$ and 0.1?

3. Which of these doesn't belong? Why?

   **a)** 0.2    **b)** $\frac{2}{10}$    **c)** $\frac{6}{100}$

   **d)** 0.4    **e)** $\frac{5}{10}$

### Booklink

**The Phantom Tollbooth** by Norton Juster and Jules Feiffer (Epstein & Carroll; distributed by Random House: New York, 1961). Milo goes on a journey through a land with words and numbers, including decimals.

## Practice

Estimate the following, then solve each question. Compare your answers with a classmate's.

| | | | |
|---|---|---|---|
| **1.** 2.10 <br> + 3.21 | **2.** 3.48 <br> + 8.99 | **3.** 4.78 <br> + 2.03 | **4.** 5.48 <br> − 2.36 |
| **5.** 8.12 <br> − 6.14 | **6.** 2.18 <br> − 1.09 | **7.** 2.03 + 9.42 = ▨ | |

**8.** 7.45 + 6.51 = ▨    **9.** 6.22 − 4.91 = ▨    **10.** 5.88 − 2.72 = ▨

Write these fractions as decimals:

**11.** $\frac{2}{5}$    **12.** $\frac{25}{100}$    **13.** $\frac{3}{4}$    **14.** $\frac{7}{8}$    **15.** $\frac{9}{10}$    **16.** $\frac{345}{1000}$    **17.** $\frac{26}{1000}$

Show these fractions using pictures:

**18.** $\frac{3}{5}$    **19.** $\frac{45}{100}$    **20.** $\frac{7}{9}$    **21.** $\frac{1}{2}$    **22.** $\frac{3}{7}$    **23.** $\frac{9}{10}$    **24.** $\frac{17}{20}$

Write these fractions on a place-value chart:

**25.** $\frac{3}{12}$    **26.** $\frac{25}{50}$    **27.** $\frac{52}{100}$    **28.** $\frac{5}{8}$    **29.** $\frac{9}{12}$    **30.** $\frac{987}{1000}$    **31.** $\frac{17}{1000}$

Solve these questions:

**32.**  3.505
  + 0.202
  _____

**33.**  5.819
  + 2.53
  _____

**34.**  4.91
  + 2.001
  _____

**35.**  3.006
  − 1.201
  _____

**36.**  5.012
  − 3.516
  _____

**37.**  2.248
  − 1.043
  _____

**38.** 3.30 + 7.742 = ■

**39.** 5.073 + 2.15 = ■   **40.** 7.778 − 6.149 = ■   **41.** 2.044 − 2.021 = ■

## Extension

Imagine you went shopping and saw a pencil for $0.29, a pen for $0.59, an eraser for $0.19, a ruler for $0.26, and a notebook for $0.75.

**42.** Calculate all the combinations you could buy with only $1.00 to spend.

**43.** What is the total amount of each combination?

### Show What You Know

**Review: Lessons 1 to 3, Operations, Fractions, and Money**

**1.** Calculate the difference in diameter between the Earth and the other planets.

**2.** Calculate the total of all the diameters of the planets and the moon.

**3. a)** If the admission to the Science Centre is $10.50 per person, how much would it cost if one class of 36 students attended? One class of 28 students? One class of 48 students?

**b)** What would be the total cost if all three classes attended? What would be the difference in cost among the three classes?

**4.** Imagine that you have three different containers filled with water. The first container holds $\frac{26}{100}$, the second container holds $\frac{38}{100}$, and the third holds $\frac{19}{100}$. If you were given a fourth container, how much water would it have to hold so that all four containers equal 1 whole?

## Lesson 4

# Solving Problems Using Operations

SPACE LOG

PLAN:
You will add, subtract, multiply, and divide numbers as you study Earth's atmosphere.

DESCRIPTION:
Between Earth and space is the atmosphere, which protects Earth like a blanket. It contains the oxygen we need to breathe along with many other gases. Beyond the atmosphere is what we call "outer space."

## Get Started

Li and Ahanu stopped to look at an exhibit about Earth's atmosphere.

They found out that the atmosphere is made up of $\frac{78}{100}$ nitrogen and $\frac{21}{100}$ oxygen. The remaining $\frac{1}{100}$ consists of many different gases.

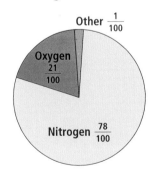

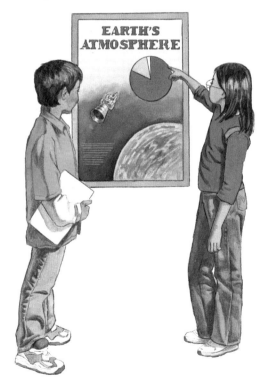

1. Convert these fractions to decimals:

   **a)** $\frac{78}{100}$     **b)** $\frac{21}{100}$     **c)** $\frac{1}{100}$

2. Record the decimals on a place-value chart like this one.

|  | Ones | Tenths | Hundredths |
|---|---|---|---|
| Nitrogen |  | . |  |
| Oxygen |  | . |  |
| Other Gases |  | . |  |

## Write Problems

Write five problems for a classmate to solve. You can use the data on Earth's atmosphere shown in the chart and graph titled "Layers of the Atmosphere" on this page, as well as the data in Get Started. You can write addition, subtraction, multiplication, and division problems.

Remember to write an answer key to your problems!

| Layers of the Atmosphere | |
|---|---|
| Name | Depth |
| Thermosphere | 120 km – 600 km |
| Mesopause | 85 km – 120 km |
| Mesosphere | 60 km – 85 km |
| Stratopause | 50 km – 60 km |
| Stratosphere | 25 km – 50 km |
| Tropopause | 15 km – 25 km |
| Troposphere | 0 km – 15 km |

**Layers of the Atmosphere**

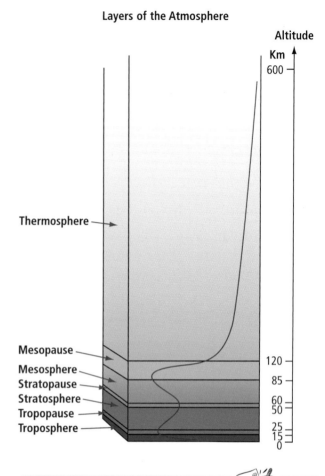

**Journal**

Write down any challenges you found when creating your problems.

Here is an example of a problem you might come up with:

The space shuttle orbits the earth for 1 week.
Here are the altitudes for each day:

Day 1 – 348 km      Day 2 – 198 km

Day 3 – 638 km      Day 4 – 550 km

Day 5 – 299 km      Day 6 – 198 km

Day 7 – 643 km

Find the mean altitude and mode for all seven days.

**Tip**

Remember, mean is the average of a set of amounts, and mode is the value that occurs most often in a set.

## What Did You Learn?

1. Were there any problems you wrote that you couldn't solve? How did you revise them?
2. Did your partner write any problems you couldn't solve? What did you do?

## Practice

Show these fractions as decimals:

**1.** $\frac{10}{100}$ **2.** $\frac{3}{5}$ **3.** $\frac{17}{20}$ **4.** $\frac{12}{15}$ **5.** $\frac{562}{1000}$ **6.** $\frac{230}{1000}$

Solve these questions:

**7.** $\begin{array}{r} 456 \\ 387 \\ 219 \\ +\ 345 \\ \hline \end{array}$ **8.** $\begin{array}{r} 362 \\ 987 \\ 103 \\ +\ 482 \\ \hline \end{array}$ **9.** $\begin{array}{r} 319 \\ 280 \\ 107 \\ +\ 999 \\ \hline \end{array}$ **10.** $\begin{array}{r} 362 \\ 100 \\ 459 \\ +\ 714 \\ \hline \end{array}$

Solve these questions:

**11.** $\begin{array}{r} 52\,312 \\ -\ 4\,623 \\ \hline \end{array}$ **12.** $\begin{array}{r} 12\,345 \\ -\ 1\,687 \\ \hline \end{array}$ **13.** $\begin{array}{r} 98\,800 \\ -\ 8\,723 \\ \hline \end{array}$

**14.** $\begin{array}{r} 80\,124 \\ -\ 6\,323 \\ \hline \end{array}$ **15.** $\begin{array}{r} 21\,803 \\ -\ 5\,189 \\ \hline \end{array}$ **16.** $\begin{array}{r} 45\,130 \\ -\ 4\,312 \\ \hline \end{array}$

Solve these questions. The first one is done for you.

**17.** $\begin{array}{r} 115 \\ 62\overline{)7134} \\ -\ 62 \\ \hline 93 \\ -\ 62 \\ \hline 314 \\ -\ 310 \\ \hline 4 \end{array}$ **18.** $36\overline{)9813}$ **19.** $14\overline{)3231}$

## Lesson 5

# Order of Operations

SPACE LOG

PLAN:
You will solve skill-testing questions in order to be chosen to go on a space mission.

DESCRIPTION:
Astronauts complete many years of schooling. Then they do more training and preparation, sometimes for years. Even then, not every astronaut is chosen to go on a space mission.

## Get Started

The astronauts who get to go on space missions are carefully chosen. In the future, it might be common for people who aren't astronauts to go to space in the way that people travel by plane today.

Suppose you could enter a contest to win a trip on a space shuttle mission with astronauts. As with most contests, you would probably have to answer a skill-testing question.

Create five skill-testing questions for a classmate to solve using the order of operations. Use the box on page 41 to help you. Here is an example:

$14 + 4 \times 20 - 8$

$= 14 + 80 - 8$

$= 94 - 8$

$= 86$

### Vocabulary

**order of operations:** The sequence in which complex equations are solved: multiplication and division first, then addition and subtraction, always working from left to right

To solve complex equations, you need to follow a specific order to get the right answer. This is called the order of operations.
Here are the rules:

- Do multiplications and divisions first, whichever comes first, working from left to right.

- Do additions and subtractions next, whichever comes first, working from left to right.

For example, for the equation

10 x 2 – 8 = ■

multiply 10 x 2 first, then subtract 8. The answer is 12.

But for the equation

10 – 2 x 4 = ■

multiply 2 x 4 first, then subtract it from 10. The answer is 2.

If you don't follow the order of operations, you won't correctly answer the skill-testing question!

## Build Your Understanding

### Order of Operations

Check your classmate's answers to the questions that you made in Get Started. If you disagree on an answer, review the order of operations.

Solve the following questions:

**1.** 265 + 104 – 52 x 8 ÷ 2

**2.** 43 x 3 + 567 ÷ 3 + 5

**3.** 65 – 3 x 18 + 2

**4.** 300 ÷ 30 – 3 + 4 x 10

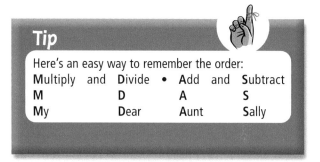

**Tip**

Here's an easy way to remember the order:

| Multiply | and | Divide | • | Add | and | Subtract |
|----------|-----|--------|---|-----|-----|----------|
| **M** | | **D** | | **A** | | **S** |
| **M**y | | **D**ear | | **A**unt | | **S**ally |

My Dear Aunt Sally

### Booklink

**Architect of the Moon** by Tim Wynne-Jones and Ian Wallace (Douglas & McIntyre: Toronto, ON, 1988, 1991). One night, the moon sends David Finebloom a message: "Help, I'm falling apart. Yours, the moon." David uses his rocket made of household objects to transport himself and his collection of blocks to the moon. Will David be able to fix the moon? Read this book to find out!

## What Did You Learn?

**1.** Why is it important to solve equations using the order of operations?

**2.** What did you find easy about solving questions using the order of operations, and what did you find most challenging?

## Practice

Solve these skill-testing questions.
Show your steps.

**1.** 54 − 3 x 12 + 2

**2.** 25 ÷ 5 x 4 − 3 x 6

**3.** 98 − 20 x 3 + 8

**4.** 12 + 9 x 7 − 4

**5.** 16 − 2 x 3 + 42

**6.** 42 ÷ 6 x 4 x 2 + 36 − 5

**7.** 33 ÷ 3 x 5 + 62

**8.** 16 x 5 ÷ 8 − 5 + 60 x 2

**9.** 20 + 30 x 5 − 16 + 4

**10.** 2 + 102 − 5 x 5 + 21

Solve these skill-testing questions:

**11.** 10 − 2 x 2 + 8 ÷ 2 x 6

**12.** 14 x 2 − 8 x 3 + 3 + 2 − 1 x 8

**13.** 33 − 4 x 8 + 1

**14.** 50 ÷ 2 + 5 x 2 ÷ 5 + 10

### Technology

Use a multimedia application to create a presentation explaining the order of operations. You might want to include some practice questions in your presentation for your audience to practise what they have learned.

## Extension

**15.** Write a problem that has two steps, which relates to space travel. Use the order of operations in your problem.

**16.** Give your problem to a classmate to solve.

**Lesson 6**

# Estimating and Calculating Products and Quotients

*SPACE LOG*

*PLAN:*
*You will find sums, work with number patterns, and estimate products and quotients for whole numbers as you solve problems at the space exhibit.*

*DESCRIPTION:*
*Sputnik, the first artificial satellite to go into space, was launched by Russia in 1957. Since then, more and more satellites are orbiting Earth. There are many other objects orbiting Earth, too, such as parts of rockets that fall away during a launch.*

## Get Started

1. What is the pattern in the group of numbers below?

   1   3   6   10

2. Now add these numbers:

   **a)** 1 + 2 = ▧

   **b)** 1 + 2 + 3 = ▧

   **c)** 1 + 2 + 3 + 4 = ▧

   **d)** 1 + 2 + 3 + 4 + 5 = ▧

   **e)** 1 + 2 + 3 + 4 + 5 + 6 = ▧

3. The numbers above are called triangular numbers. Write down the next five triangular numbers.

4. What is the pattern in these numbers?

   1   4   9   16

### Vocabulary

**addend:** Any number that is added
**product:** The answer to a multiplication question
**quotient:** The answer to a division question
**sum:** The answer to an addition question
**triangular number:** A whole number that can be represented by dots shown in a triangular array; for example, 1, 3, and 6 are the first three triangular numbers.

**5.** Now add these numbers:

   **a)** 1 + 3 = ■

   **b)** 3 + 6 = ■

   **c)** 6 + 10 = ■

   **d)** 10 + 15 = ■

   **e)** 15 + 21 = ■

**6.** What is the pattern in the addition questions above? What do you notice about the addends? What do you notice about the sums?

**7.** The sums in question 5 are called square numbers. Why? What are the next five sums in the pattern?

## Build Your Understanding

### Estimate Multiplication and Division

At the space exhibit, Eric and Claire learned that almost 4800 satellites had been launched into space since the first Russian satellite, Sputnik. They also found out that almost 2300 of them are still in orbit, even though some are no longer in use.

"Let's see," said Eric. "If the first satellite went up in 1957, and almost 5000 satellites have been launched altogether, what's the average number of satellites launched every year?"

"That's easy," said Claire. "Around 100."

**1.** Sometimes, it is easier to estimate the answer to a question by rounding the numbers. For example, it is easier mentally to work with 5000 and 50 instead of 4800 and 46. Look at the illustration to the right and examine Claire's solving method. Now check her work.

Hmmm. I'll round 1957 up to 1960 and I'll round 2003 down to 2000. 2000 - 1960 = 40. Since 1960 - 1957 = 3, and 2003 - 2000 = 3, I will add 3 + 3 + 40 = 46. Now I'll round 46 up to 50. 5000 ÷ 50 = 100. That's about 100 satellites per year.

**2.** Estimate the solutions to these questions:

| | | | | |
|---|---|---|---|---|
| **a)** 20 <br> x 45 | **b)** 78 <br> x 42 | **c)** 14 <br> x 8 | **d)** 80 <br> x 12 | **e)** 54 <br> x 13 |
| **f)** 87 <br> x 45 | **g)** 10 <br> x 23 | **h)** 11 <br> x 23 | **i)** 46 <br> x 45 | **j)** 59 <br> x 6 |

**k)** $5{\overline{)54}}$   **l)** $3{\overline{)100}}$   **m)** $20{\overline{)45}}$   **n)** $300 \div 45$   **o)** $202 \div 15$

**3.** Now check your estimates by solving each question.

**Journal**

Choose a problem for which you estimated a solution. Write down the steps you took mentally as you worked through the problems.

## What Did You Learn?

**1.** Explain what a pattern is, using numbers, pictures, and words.

**2.** In your own words, explain square numbers to a classmate.

**3.** Explain how estimating first can help you solve a problem.

## Practice

**1.** Work with a classmate and brainstorm a list about patterns. You may want to think about the following questions:
- What makes a pattern?
- Where are patterns found?
- How can you use addition, subtraction, multiplication, and division to show patterns?

Estimate the solutions to these questions. Then solve each question to check your estimate:

| | | | | |
|---|---|---|---|---|
| **2.** 45 <br> x 34 | **3.** 86 <br> x 12 | **4.** 23 <br> x 35 | **5.** 89 <br> x 11 | **6.** 43 <br> x 21 |
| **7.** 98 <br> x 32 | **8.** 67 <br> x 76 | **9.** 45 <br> x 33 | **10.** 61 <br> x 32 | |

**11.** $56 \div 3 = $ ■    **12.** $25\overline{)98}$    **13.** $24\overline{)56}$    **14.** $23\overline{)99}$

**15.** $78 \div 12 = $ ■    **16.** $96 \div 23 = $ ■    **17.** $76 \div 15 = $ ■

**18.** $13\overline{)73}$    **19.** $89 \div 32 = $ ■    **20.** $29\overline{)57}$

Extend the following patterns:

**21.** 34, 40, 37, 43, 40, 46, 43, 49, ■, ■, ■

**22.** 118, 119, 128, 129, 138, 139, 148, 149, ■, ■, ■

**23.** 45, 51, 58, 66, 75, ■, ■, ■

## A Math Problem to Solve

**24.** Imagine that there is a row of 27 satellite dishes in a straight line. Each satellite dish has a diameter of 25 m, and there is an equal number of space between each satellite dish. If the entire structure is 3 km across, how much space is between each satellite dish? Explain the steps you followed to solve this problem.

**Lesson 7**

# Estimating Products and Quotients and Analyzing Data

*SPACE LOG*

*PLAN:*
*You will estimate the answers for multiplication and division problems with decimals and fractions as you analyze space shuttle size, number of astronauts, and days in space. You will also calculate mode and mean.*

*DESCRIPTION:*
*The space shuttle is what NASA uses to take astronauts to space and back. It's also where the astronauts eat, work, and sleep for the entire mission. A mission can be anywhere from 5 to 16 days long. A mission crew can include as many as ten people.*

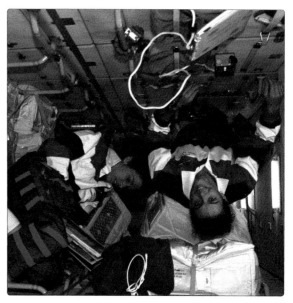

**Canadian Astronaut Julie Payette (right)**

## Get Started

The orbiter is the part of the space shuttle where the astronauts live while in space. It is 56.14 m long.

1. If there are 7 astronauts on board and they were to lie down end to end, how many metres of space would each astronaut have? Estimate the answer.

2. About how many metres would each astronaut have if there were 6 crew members on board? Estimate the answer.

3. Now solve these problems. Check if your estimates were close.

### Tip

When you divide a decimal number by a one-digit whole number, remember to line up the place values in your answer with the correct place values in the dividend.

The answer is 6.24, remainder 7.

$$
\begin{array}{r}
6.24 \\
9\overline{)56.23} \\
-54 \\
\hline
2\ 2 \\
-1\ 8 \\
\hline
4\ 3 \\
-3\ 6 \\
\hline
7
\end{array}
$$

### Vocabulary

**dividend:** The number being divided
**divisor:** The number that divides the dividend

## Estimate Fractions and Decimals

The space shuttle *Discovery* was launched on October 11, 2000, which marked the hundredth space shuttle mission.

1. On 100 space missions, $\frac{25}{100}$ were each 6 days long, $\frac{36}{100}$ were each 10 days long, $\frac{4}{100}$ were each 5 days long, $\frac{18}{100}$ were each 16 days long, and the rest were each 12 days long.

   **a)** What fraction of the missions were 12 days long? Convert this fraction to a decimal.

   **b)** What is the mode trip length in days? How do you know?

   **c)** What fraction of the missions were less than 10 days long? Convert this fraction to a decimal.

   **d)** What fraction of the missions were at least 10 days long? Convert this fraction to a decimal.

   **e)** Estimate the number of days that were spent in space during the 100 missions. Check your estimate.

   **f)** Look carefully at how many days each mission took. Estimate how many days an average (mean) space mission would be. Explain your thinking.

2. The average crew is 5 to 7 people.

   **a)** If $\frac{25}{100}$ missions each had 5 crew, $\frac{48}{100}$ missions each had 6 crew, and the rest had 7 crew, how many astronauts went in total?

   **b)** What is the mode number of crew members on a space mission? Explain your answer.

### Vocabulary

**mean:** The average of a set of amounts. To calculate the mean, add up all the amounts and divide the sum by the total number of addends given. The mean of 2, 3, 5, 6, and 9 is 5.
**mode:** The value that occurs most often in a set. The mode of 2, 3, 5, 5, 7, 5, 2, 8, 5 is 5.

**c)** If there were 200 space missions, how many crew members would you expect to participate?

**d)** If 100 new space missions are planned and 0.25 of these missions only need 3 crew members, how many missions will need 3 crew members?

## What Did You Learn?

**1.** Explain the steps you followed to find the mean and the mode.

**2.** How close were your estimates to the solutions?

**3.** How can you improve your estimates?

**4.** Explain how you converted fractions to decimals.

## Practice

Estimate the following:

| **1.** | 6.345 | **2.** | 3.497 | **3.** | 6.424 | **4.** | 7.012 |
|---|---|---|---|---|---|---|---|
| | x 2 | | x 3 | | x 5 | | x 4 |

| **5.** | 9.326 | **6.** | 8.001 | **7.** | 6.132 | **8.** | 3.125 |
|---|---|---|---|---|---|---|---|
| | x 3 | | x 2 | | x 4 | | x 5 |

Estimate the following:

**9.** $56.23 \div 8 =$ ▨

**10.** $42.35 \div 7 =$ ▨

**11.** $32.013 \div 8 =$ ▨

**12.** $99.03 \div 3 =$ ▨

**13.** $12.423 \div 3 =$ ▨

**14.** $80.99 \div 9 =$ ▨

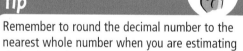

**Tip**

Remember to round the decimal number to the nearest whole number when you are estimating decimal numbers.

Find the mode of these numbers:

**15.** 36, 27, 32, 48, 27, 92, 36, 27, 103

**16.** 362, 389, 411, 516, 380, 516, 230, 193, 516

**17.** 179, 125, 109, 170, 109, 125, 117, 109, 198, 182

Find the mean of these numbers:

**18.** 62, 38, 47, 59, 30, 42

**19.** 562, 871, 602, 598, 437, 623

**20.** 890, 999, 623, 711, 892, 765

## A Math Problem to Solve

**21.** Make up your own problems based on the information in Build Your Understanding. Trade problems with a classmate. Estimate first, then check your solutions and compare them to your estimates.

### Booklink

**On the Shuttle: Eight Days in Space** by Barbara Bondar and Roberta Lynn Bondar (Greey de Pencier Books: Toronto, ON, 1993). This non-fiction book describes Roberta Bondar's space shuttle mission and the experiments she participated in. She was the first Canadian woman in space.

## Show What You Know

### Review: Lessons 4 to 7, Order of Operations, Products, and Quotients

**1.** Use the order of operations to make five different equations that equal 20.

**2.** Explain the difference between triangular numbers and square numbers using numbers, pictures, and words.

**3. a)** If an astronaut had to be trained for 3 years, how many months was she training for? How many weeks was she training for?

**b)** If an astronaut trained for 5 days a week for 10 hours each day, how much training would he receive in 1 week? 1 month? 1 year? 3 years?

**Lesson 8**

# Multiplying Three-Digit Numbers by Two-Digit Numbers

SPACE LOG

PLAN:

You will multiply three-digit numbers by two-digit numbers, and add and subtract money amounts to plan and calculate costs for a shuttle mission menu.

DESCRIPTION:

Planning the food for each shuttle mission takes a great deal of work. NASA chooses menus that meet all the astronauts' daily nutritional requirements. The astronauts do get to have some say in what they eat, but they make their choices five months before their mission!

## Get Started

1. One person needs about 2 L of water a day. If a mission has 5 crew members and is going to last for 14 days, how much water is needed?

2. One person needs about 500 g of dry food a day. The mission will last for 12 days. How much food is needed per person?

| Here is one method to help you. | Then multiply the tens. | Then add. |
|---|---|---|
| First multiply the ones. | 500 | 500 |
| 500 | x    12 | x    12 |
| x    12 | 1000 | 1000 |
| 1000 | 5000 | + 5000 |
|  |  | 6000 |

## Multiply Menus

The next space shuttle mission is preparing its grocery list. There are 7 crew members. The mission is being planned for 15 days, but an extra 3 days of supplies are taken along just in case there are problems and the shuttle can't return to Earth on schedule.

The possible menu items and their prices are shown below:

| Meal 1 | Meal 2 | Meal 3 | Meal 4 | Meal 5 |
|---|---|---|---|---|
| Dried Peaches | Ham | Chicken à la King | Dried Pears | Salmon |
| Cornflakes | Cheese Spread | Turkey Tetrazzini | Sausage Patty | Tortilla |
| Orange-Pineapple Drink | Tortilla | Cauliflower with Cheese | Mexican Scrambled Eggs | Pears |
| Cocoa | Pineapple | Brownie | Grits with Butter | Chocolate Cookies |
| $10.00 | Cashews | Grape Drink | Orange Juice | Lemonade |
| | Strawberry Drink | $25.00 | $15.00 | $20.00 |
| | $15.00 | | | |

| Meal 6 | Meal 7 | Meal 8 | Meal 9 | Meal 10 |
|---|---|---|---|---|
| Beef Tips with Mushrooms | Dried Pears | Peanut Butter | Frankfurters | Dried Apricots |
| Noodles & Chicken | Beef Patties | Apple or Grape Jelly | Macaroni & Cheese | Granola with Blueberries |
| Creamed Spinach | Scrambled Eggs | Tortilla | Green Beans with Mushrooms | Orange-Grapefruit Drink |
| Peaches | Vanilla Instant Breakfast | Fruit Cocktail | Peach Ambrosia | $10.00 |
| Granola Bar | Orange Juice | Trail Mix | Tropical Punch | |
| Tea with Lemon | $15.00 | Peach-Apricot Drink | $20.00 | |
| $25.00 | | $10.00 | | |

| Meal 11 | Meal 12 | Meal 13 | Meal 14 | Meal 15 |
|---|---|---|---|---|
| Chicken Salad Spread | Sweet 'n' Sour Beef | Dried Apricots | Turkey Salad Spread | Spaghetti with Meat Sauce |
| Crackers | Rice Pilaf | Breakfast Roll | Tortilla | Italian Vegetables |
| Chocolate Pudding | Broccoli au Gratin | Chocolate Instant Breakfast | Peaches | Butterscotch Pudding |
| Butter Cookies | Vanilla Pudding | Grapefruit Juice | Granola Bar | Orange Drink |
| Lemonade | Apple Cider | $10.00 | Lemonade | $20.00 |
| $15.00 | $25.00 | | $15.00 | |

| Meal 16 | Meal 17 | Meal 18 | Meal 19 | Meal 20 |
|---|---|---|---|---|
| Dried Pears | Tuna | Shrimp Cocktail | Dried Peaches | Dried Beef |
| Beef Patty | Tortilla | Beef Steak | Bran Cereal | Cheese Spread |
| Scrambled Eggs | Banana Pudding | Potatoes au Gratin | Orange-Mango Drink | Applesauce |
| Oatmeal with Brown Sugar | Shortbread Cookies | Asparagus | Cocoa | Peanuts |
| Orange Juice | Almonds | Strawberries | $10.00 | Tropical Punch |
| $15.00 | Grape Drink | Lemonade | | $15.00 |
| | $15.00 | $25.00 | | |

Plan a menu for the mission. All 7 crew members will have the same breakfasts, lunches, and dinners each day. A single meal may be repeated throughout the 15 days, but your combination of three meals should be different every day. The entire food supply for the mission cannot go over the budget of $7000.00.

## What Did You Learn?

How did you solve this problem? Did you plan the meals first and then calculate the costs? Or was it easier to set a daily budget and choose the meals afterward? Explain your choices.

## Practice

Solve these problems:

| | | | | | | | |
|---|---|---|---|---|---|---|---|
| **1.** | 101<br>x 32 | **2.** | 283<br>x 16 | **3.** | 573<br>x 51 | **4.** | 499<br>x 24 |
| **5.** | 982<br>x 97 | **6.** | 387<br>x 41 | **7.** | 874<br>x 11 | **8.** | 259<br>x 13 |

**9.** Review the steps you used to plan the meals for your space mission in this lesson's Build Your Understanding.

**10.** Compare your steps with a classmate's. How are they the same, and how are they different?

**Technology**

Use a computer program to create a personal math reference guide. Think about all the math you have learned so far and record instructions, rules to follow, and strategies that will help you. You can keep adding to your reference guide as you continue the school year.

**Booklink**

**Math Curse** by Jon Scieszka and Lane Smith (Viking: New York, 1995). One student finds that math really is everywhere—so much so that it feels like it has become a curse! But in the end, this student learns that math is both useful and fun!

## Lesson 9

# Multiples of 2 to 10

SPACE LOG

PLAN:
You will find multiples of a variety of numbers and learn how to factor during your time on the space mission.

DESCRIPTION:
Life on the space shuttle while it is orbiting Earth can become pretty routine. The astronauts spend their time doing science experiments, performing maintenance on satellites, or building the space station. They also eat, sleep, and exercise. Believe it or not, all of these activities use math!

## Get Started

Use the tips on the next page to answer these questions.

**1.** Which of these numbers are multiples of 2?

8, 9, 11, 24, 4, 17, 16, 10, 55

**2.** Which of these numbers are multiples of 3?

6, 33, 14, 18, 7, 22, 27, 97, 81

**3.** Which of these numbers are multiples of 5?

25, 44, 100, 12, 20, 82, 85, 99, 20

**4.** In your notebook, write down one example for each of the tips on the next page.

### Vocabulary

**multiple:** The product of a given whole number and another whole number other than 1. For example, some multiples of 6 are 6, 12, 18, 24, etc.

### Journal

Write down how you found the answers to the questions in Get Started.

**5.** Compare your examples with a classmate's, and check his or her work.

## Tips for Determining Multiples

- All whole numbers can be divided by 1.
- All even numbers can be divided by 2.
- If the sum of all the digits in a number can be divided by 3, then the number can be divided by 3.
- If the last 2 digits in a number make a number that can be divided by 4, then the entire number can be divided by 4.
- If the number ends in 0 or 5, it can be divided by 5.
- If a number can be divided by both 2 and 3, then it can be divided by 6.
- If the last 3 digits in a number make a number that can be divided by 8, then the entire number can be divided by 8.
- If the sum of all the digits in a number can be divided by 9, then the number can be divided by 9.
- If a number ends in 0, then it can be divided by 10.

## Build Your Understanding

### Factor on the Shuttle

The astronauts have 24 experiments to perform on their mission. They want to do an equal number of experiments each day. They will decide how long the mission will be when they know how many experiments they can do each day. How can they calculate this information?

First, decide how to solve the problem.

- You can divide 24 by the number of experiments you think the astronauts might be able to do in a day.

24 ÷ 1 = ▦

24 ÷ 2 = ▦

24 ÷ 3 = ▦

24 ÷ 4 = ▦

24 ÷ 5 = ▦

24 ÷ 6 = ▦

24 ÷ 7 = ▦

- You can also use objects and divide them into groups until you have groups of equal sizes.
- You can also write multiplication statements that have a product of 24.

  1 x 24 = 24

  2 x 12 = 24

  3 x 8 = 24

  4 x 6 = 24

The numbers in these multiplication statements are called factors.

1, 2, 3, 4, 6, 8, 12, and 24 are factors of 24.

## What Did You Learn?

**1.** How many experiments should be done each day if the mission has 6 days available for doing the experiments? How many if there are 12 days available?

**2.** Which approach to solving this problem did you find easiest? Which approach did you find most challenging? Explain your answers.

## Practice

**1.** List all of the factors for these numbers:

   3    12    36    25    48    51    63

**2.** Write these numbers in your notebook. Circle the multiples of 3 in red. Circle the multiples of 5 in green.

   3, 5, 8, 9, 12, 15, 16, 18, 22, 25, 33, 35, 40, 42

**3.** List all the multiples of 5 up to 100.

**4.** List all the multiples of 10 up to 200.

**5.** List all the multiples of 3 up to 50.

## Lesson 10

# Multiplying and Dividing Decimals

SPACE LOG

PLAN:
You will multiply and divide decimals, including money amounts, to calculate supplies for a mission at the International Space Station.

DESCRIPTION:
The International Space Station will allow astronauts to live and work in space for months at a time. But a lot of supplies will be needed to support the people who live at the station.

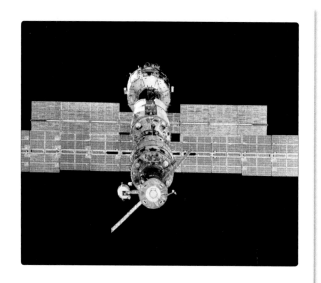

## Get Started

Ajay, Phillip, and Li wondered how all the materials needed to complete the International Space Station get to the space station. They found out that over 40 space flights by NASA's space shuttle and Russian rockets will deliver all the needed components.

Space shuttle *Discovery* can carry 900 kg and space shuttle *Atlantis* can carry 750 kg of materials. If *Discovery* makes 20 trips to the International Space Station, how many trips would *Atlantis* have to make if NASA had to deliver 36 000 kg of supplies?

### Technology

Imagine that your school has decided to sell sportswear as a fundraiser. Work with a partner and decide on what merchandise should be included in the fundraiser and for what price. Design and create a brochure using a desktop publishing program to advertise the merchandise. Distribute your brochure to others in the class, who can then place orders for the merchandise. What is the total amount of your sales, including taxes?

### Supply the Station

Here is the list of supplies needed for the space station.

- The cost of food per astronaut per day is $54.08.

- Each astronaut needs the following clothing and personal supplies for his or her trip:

| Item | Cost per item |
|---|---|
| 2 pairs of pants | $62.99 each pair |
| 5 T-shirts | $9.15 each |
| 5 pairs of socks | $2.41 each pair |
| Toothbrush | $3.75 |
| Comb | $4.50 |
| 2 sweaters | $35.34 each |

- The astronauts will conduct experiments 5 days out of every 7 days for 56 days. They conduct one experiment each day. Each experiment requires supplies that cost an average of $125.75.

There will be 3 astronauts living on the station for 56 days. What is the total cost of the clothing, personal items, and experiment supplies needed for all astronauts? You can use a calculator if it would be helpful.

## What Did You Learn?

1. In your notebook, record the steps you used to calculate the total cost of staying at the space station.

2. How did you calculate the total cost of experiment supplies? List the steps you followed.

3. Compare your answers in questions 1 and 2 with a classmate's. How are they the same and how are they different?

**1.** 0.785
x 4

**2.** 0.693
x 8

**3.** 0.105
x 9

**4.** 0.243
x 2

**5.** 1.531
x 2

**6.** 0.148
x 3

**7.** 0.582
x 3

**8.** 0.811
x 3

**9.** 2.568 ÷ 3 = ▦

**10.** 3.36 ÷ 4 = ▦

**11.** 1.772 ÷ 2 = ▦

**12.** 2.55 ÷ 5 = ▦

**13.** 6)‾1.386

**14.** 8)‾0.904

**15.** 3)‾0.999

**16.** 4)‾0.464

## Math Problems to Solve

**17.** If the total budget for food during a 30-day space mission is $8263.00, how many astronauts can go on the mission?

**18.** Will there be any money left over?

### Show What You Know
**Review: Lessons 8 to 10, Multiplying and Dividing**

**1. a)** Imagine that a restaurant was given a budget of $200.00 for one week to plan meals for a group of 4 people. If the restaurant spent an equal amount each day, what would the cost be each day?

**b)** If you were calculating three meals per day, what would the approximate cost be for breakfast? lunch? dinner?

**c)** Compare your answers with a classmate's, and then share them with the class.

**2.** Imagine that the astronauts on one mission have to perform 36 experiments. How long might their mission be if they want to perform an equal number of experiments each day? Show all possible answers. Which of your answers makes the most sense? Why?

**3.** While you were shopping, you saw:

**a)** a pair of jeans for $39.99

**b)** a T-shirt for $12.50

**c)** running shoes for $64.30

**d)** a sweatshirt for $32.50

**e)** socks for $3.99 a pair

If you had a gift certificate for $150.00, what combinations could you buy, not going over $150.00? Don't forget to calculate the GST in your costs.

## Lesson 11

# Estimating and Calculating Products and Quotients

SPACE LOG

PLAN:
You will estimate and then calculate the products and quotients of decimals.

DESCRIPTION:
Estimating numbers can help you quickly calculate the answers to math questions. You might have to quickly figure out how much money you will need to spend on supplies, or how much space there will be in the space shuttle.

## Get Started

1. If there is 11.88 m² of area in one part of the space shuttle, and you need to share this area evenly with 3 astronauts, how can you quickly figure this out?

   If we round 11.88 to the nearest whole number, it is 12.

   We know that 12 ÷ 3 = 4.

   Therefore, each astronaut will have approximately 4 m² of area.

2. If there are 35.56 packages of food for the space mission each day, and there are 12 astronauts who each need 3 packages a day, will there be enough food for the astronauts? How do you know?

## Estimate and Divide with Decimals

**1.** Estimate the answers to these questions:

    **a)** 11.88 ÷ 3 = ▓      **b)** 32.12 ÷ 2 = ▓

    **c)** 23.78 ÷ 4 = ▓      **d)** 24.67 ÷ 5 = ▓

    **e)** 3.23 x 5 = ▓       **f)** 50.98 x 3 = ▓

    **g)** 99.76 x 4 = ▓      **h)** 2.34 x 16 = ▓

**2.** Find the actual answers to the questions above.

**3.** What is the difference between the actual answers and your estimates?

**Tip**

Remember to round the decimal number to the nearest whole number when you are estimating decimal numbers.

## What Did You Learn?

**1.** List the steps you completed to solve question 2 in Get Started.

**2.** Explain how you estimated the answers to the questions in this lesson's Build Your Understanding.

**3.** Compare your method of estimating products and quotients to a classmate's.

**4.** Explain how rounding numbers helps you estimate.

## Practice

Mentally estimate these division equations:

**1.** 36.317 ÷ 2 = ■ **2.** 45.036 ÷ 3 = ■

**3.** 16.02 ÷ 4 = ■ **4.** 80.987 ÷ 9 = ■

Mentally estimate these multiplication equations:

**5.** 2.45 x 15 = ■ **6.** 5.78 x 8 = ■

**7.** 24.061 x 2 = ■ **8.** 49.55 x 5 = ■

**9.** Find the actual answers to the questions above.

**10.** How do your estimates compare to your actual answers? What is the difference for each?

### A Math Problem to Solve

**11.** A day on Mars is 24 h, 35 min, and 23 s long. How much longer is this than one Earth day? If you lived on Mars, how old would you be in Mars days?

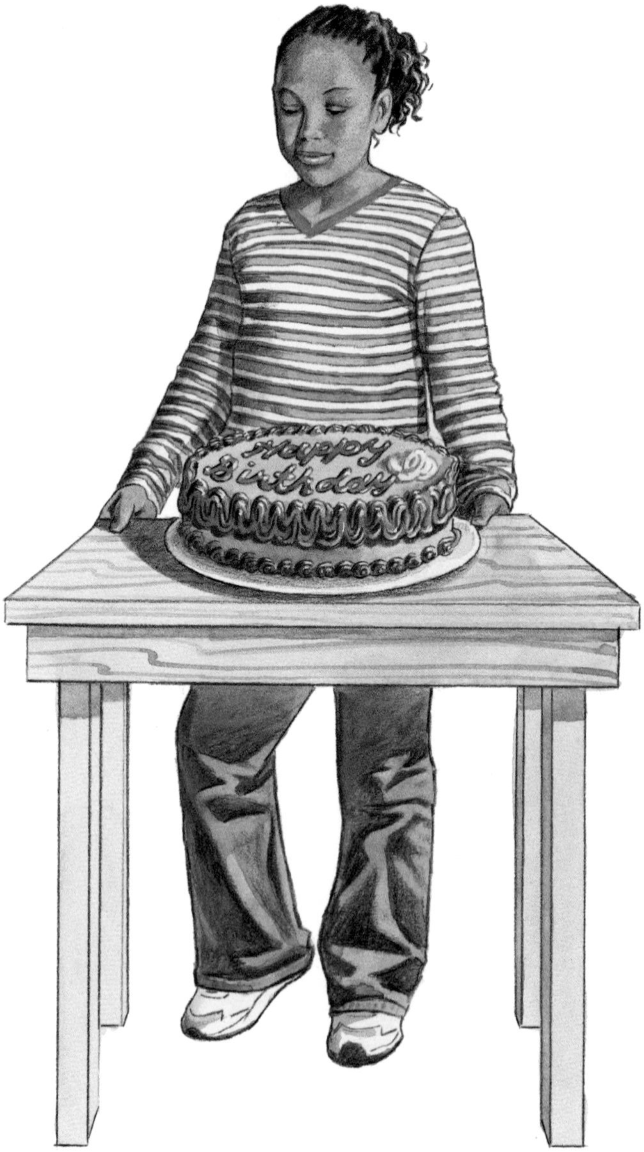

# Chapter Review

**1.** Add the following:

**a)**  1328
  + 1976

**b)**  6105
  + 3225

**c)**    530
    280
    896
  + 403

**d)** 382 + 197 + 398 + 891 = ■

**e)** 356 + 107 + 235 + 987 = ■

**2.** Subtract the following:

**a)**  2315
  − 2264

**b)**  5027
  − 2130

**c)** 54 130
  − 2 260

**d)** 63 122 − 4001 = ■

**e)** 82 438 − 2113 = ■

**3.** Multiply the following:

**a)**    99
  x 22

**b)**    31
  x 11

**c)**  126
  x 54

**d)** 387 x 46 = ■

**e)** 262 x 41 = ■

**4.** Divide the following:

**a)** 162 ÷ 18 = ■

**b)** 408 ÷ 17 = ■

**c)** 294 ÷ 42 = ■

**d)** 15)1245

**e)** 35)3150

**f)** 14)3972

**5.** Add the following:

**a)**  1.981
  + 0.406

**b)**  1.375
  + 1.27

**c)**  1.49
  + 1.201

**d)**    2.03
  + 4.125

**e)** 4.579 + 1.96 = ■

**6.** Subtract the following:

**a)** 3.876
  − 1.304

**b)** 1.708
  − 1.609

**c)** 3.907 − 2.361 = ■

**d)** 9.882
  − 4.907

**e)** 6.504
  − 6.001

**7.** Make a place-value chart in your notebook, ten thousands to thousandths. Place these numbers on your chart:

**a)** 756.01   **b)** $\frac{60}{100}$   **c)** $\frac{9}{10}$   **d)** 10 251   **e)** 1300

**f)** six hundredths   **g)** $\frac{32}{1000}$   **h)** 35.019

**8.** Identify the place value of the underlined digit.

**a)** 47<u>5</u>0   **b)** 1<u>3</u> 200   **c)** 812.<u>1</u>7

**d)** 7<u>2</u>.800   **e)** 7.5<u>2</u>   **f)** 10.95<u>4</u>

**9.** Find the solutions to these problems. Show your steps.

**a)** 80 ÷ 5 x 8 − 8 x 2   **b)** 79 − 2 x 24 + 9

**c)** 21 + 6 x 9 − 48   **d)** 101 − 11 x 4 + 7

**10.** Estimate then calculate these division questions:

**a)** 3.136 ÷ 3 = ■   **b)** 4.319 ÷ 2 = ■

**c)** 7$\overline{)14.345}$   **d)** 9$\overline{)27.339}$

**e)** 6$\overline{)18.316}$

**11.** Estimate these multiplication questions:

**a)** 1.423
  x 3

**b)** 5.594
  x 4

**c)** 9.157
  x 6

**d)** 7.604
  x 5

**e)** 6.334
  x 9

**12.** Imagine you were going furniture shopping and saw

    **a)** a lamp for $16.99

    **b)** a coffee table for $50.25

    **c)** a chair for $39.95

    **d)** a filing cabinet for $40.15

    **e)** storage boxes for $5.99 each

If you had $150.00 to spend, what combinations could you buy without going over $150.00. Don't forget to calculate taxes into your costs.

**13.** List five multiples for these numbers:

    **a)** 8   **b)** 5

**14.** List all of the factors for these numbers:

    **a)** 9      **b)** 21     **c)** 54     **d)** 50

    **e)** 72    **f)** 79    **g)** 101

**15.** Multiply the following:

    **a)** $\begin{array}{r} 0.386 \\ \times\ 3 \\ \hline \end{array}$    **b)** $\begin{array}{r} 2.56 \\ \times\ 7 \\ \hline \end{array}$    **c)** $\begin{array}{r} 1.289 \\ \times\ 6 \\ \hline \end{array}$

    **d)** $\begin{array}{r} 0.349 \\ \times\ 8 \\ \hline \end{array}$    **e)** $\begin{array}{r} 0.107 \\ \times\ 3 \\ \hline \end{array}$    **f)** $\begin{array}{r} 0.813 \\ \times\ 5 \\ \hline \end{array}$

**16.** Divide the following:

    **a)** $6\overline{)4.464}$    **b)** $7\overline{)0.847}$    **c)** $5\overline{)5.475}$

    **d)** $3\overline{)9.374}$    **e)** $4\overline{)0.464}$    **f)** $9\overline{)8.739}$

# Chapter Wrap-Up

You have reached the end of Chapter 1. In this chapter, you added, subtracted, multiplied, and divided large and small numbers. You improved your estimating skills, and you practised finding multiples and factors of numbers.

In this chapter project, you will plan a new mission to the moon. Your astronauts will travel on the space shuttle. You need to set a budget, decide how many people will go on the mission, choose the number of experiments they will conduct on the moon, and supply your mission while meeting your budget. You will also want to alert the media about the upcoming mission. What math skills will you need to use for this project?

## Brainstorm and Review

Work in small groups and review all of the math you have learned in this chapter. Think about how you might apply it to your own space mission.

### Technology

Use an electronic graphic organizer to create a newspaper report of topics, skills, and concepts you need to consider to prepare for the space mission. Remember to include appropriate pictures and charts.

## Work on Your Project

On your own, you need to do the following:

- Think of some skill-testing questions to help you choose which of your astronauts will take part in the space mission. You will also need to decide on the number of astronauts who will take part.

- Determine how large the space shuttle will be, and calculate how many metres of lying-down space each of your astronauts will have. You can use the information in Lesson 7 to help you.

- Decide how long your space mission will be.

- Prepare your grocery list based on the information in Lesson 8 for your number of astronauts and the number of days of your space mission.
- Decide how many experiments the astronauts will do in total so they can do an equal number each day.
- Calculate the total cost of experiment supplies, clothing and personal items, and cost of food for each astronaut each day, based on the information in Lesson 10.
- Prepare a newspaper article to let people know about your exciting space mission! Be sure to include all of your calculations.

## Present Your Project

Decide on a way to present your project to the rest of your class.

Good luck!

# Chapter 2

# Math From the Moon to Mars

You will use math to learn more about the two closest objects to Earth — the moon and Mars.

In this chapter, you will

- study patterns and their rules
- learn more about multiples
- practise factoring
- learn how to solve for a missing term
- find prime numbers
- find out how decimals and percents are related
- multiply and divide by decimals

At the end of this chapter, you will design a solar system game using all the math skills you have learned.

Answer these questions to prepare for your adventure:

1. Give an example of a pattern, and explain why it is a pattern.
2. In your own words, write a definition of multiples and compare it with a classmate's.
3. Explain how decimals and fractions are related. How are they the same, and how are they different?

### Lesson 1

# Exploring Number Patterns

*SPACE LOG*

*PLAN:*
**You will identify, extend, and create patterns.**

*DESCRIPTION:*
The moon is the most visible object in our sky. It is the only place beyond Earth where humans have gone. The moon takes exactly 27.3 Earth days to revolve around Earth. It also takes exactly 27.3 Earth days to rotate on its axis. These are all patterns.

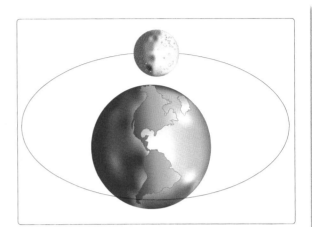

## Get Started

**1.** What is this pattern?
24, 48, 72, 96, 120

**2.** What is this pattern?
27.3, 54.6, 81.9, 109.2, 136.5

The second pattern is based on the number of Earth days it takes the moon to complete one rotation on its axis. One day on the moon is 27.3 Earth days!

## Journal

Describe what it would be like to live on "moon time." In one Earth month, how often would you be able to watch a sunset?

## Identify Patterns

**1.** What are the next five numbers for each of these patterns? Explain how you know.

**a)** 1, 2, 3, 4, ■, ■, ■, ■, ■

**b)** 2, 4, 6, 8, ■, ■, ■, ■, ■

**c)** 3, 6, 9, ■, ■, ■, ■, ■

**d)** 1, 2, 4, 8, 16, ■, ■, ■, ■, ■

**e)** 1, 3, 9, 27, ■, ■, ■, ■, ■

**f)** 1, 4, 16, 64, ■, ■, ■, ■, ■

**g)** 2, 8, 7, 13, 12, 18, 17, ■, ■, ■, ■, ■

**h)** 2, 4, 6, 2, 4, 6, ■, ■, ■, ■, ■

**i)** ■, ▲, ●, ■, ▲, ●, ■, ■, ■, ■, ■

**j)** ⠒, ⠲, ⠒⠒, ⠒⠒⠒, ⠲⠒, ⠲⠲, ⠒⠲, ⠒⠒⠲, ⠲⠒⠒, ■, ■, ■, ■, ■

**2.** Create your own patterns using numbers or shapes. Write the first four numbers or shapes of these patterns, and give them to a classmate to complete. Discuss the rule for each pattern.

## What Did You Learn?

**1.** What makes a pattern a pattern?

**2.** What is the pattern to this set of numbers?

    1   10   2   44   5   6   19

**3.** Do all sets of numbers have a pattern? If you roll a number cube 10 times in a row, how do the 10 numbers make a pattern?

1. Begin more patterns of your own, and give them to a classmate to complete. Challenge each other to make a pattern no one in your class can figure out! Be prepared to explain your pattern rule.

Copy and complete these patterns and explain the rule for each:

2. 64, 70, 68, 74, 72, 78, ■, ■, ■, ■

3. 2, 102, 4, 104, 6, 106, ■, ■, ■, ■

4. 5, 15, 10, 20, 15, 25, 20, ■, ■, ■, ■

## Extension

5. Copy a hundreds chart like the one below into your notebook, but do not shade any squares. Then, make patterns by shading numbers on the chart. For example, the shaded numbers on the chart below forms a pattern. What is the pattern? Try to think of a different pattern, and write down the numbers you shaded.

| 1 | 2 | 3 | 4 | 5 | 6 | 7 | 8 | 9 | 10 |
| 11 | 12 | 13 | 14 | 15 | 16 | 17 | 18 | 19 | 20 |
| 21 | 22 | 23 | 24 | 25 | 26 | 27 | 28 | 29 | 30 |
| 31 | 32 | 33 | 34 | 35 | 36 | 37 | 38 | 39 | 40 |
| 41 | 42 | 43 | 44 | 45 | 46 | 47 | 48 | 49 | 50 |
| 51 | 52 | 53 | 54 | 55 | 56 | 57 | 58 | 59 | 60 |
| 61 | 62 | 63 | 64 | 65 | 66 | 67 | 68 | 69 | 70 |
| 71 | 72 | 73 | 74 | 75 | 76 | 77 | 78 | 79 | 80 |
| 81 | 82 | 83 | 84 | 85 | 86 | 87 | 88 | 89 | 90 |
| 91 | 92 | 93 | 94 | 95 | 96 | 97 | 98 | 99 | 100 |

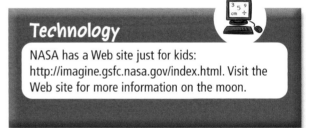

**Technology**

NASA has a Web site just for kids: http://imagine.gsfc.nasa.gov/index.html. Visit the Web site for more information on the moon.

**Lesson 2**

# Exploring More Number Patterns

SPACE LOG

PLAN:
You will look for more patterns in moon phases.

DESCRIPTION:
Every few days the moon appears to change shape because, for part of each month, it is hidden by Earth's shadow. These shapes are called the moon phases. The moon takes 29 days to go through all its phases.

## Get Started

**1.** What are the next five numbers in this pattern?

31, 28, 31, 30, 31, 30, 31, 31, 30, 31, 30, 31, 31, 28

**Tip**

There are 12 numbers in the pattern before it starts to repeat.

You Will Need
• calendar

Li and Phillip learn from the moon display at the Science Centre that the moon takes 29 days to revolve around Earth.

2. If a scientist begins tracking the moon's orbit around the earth in January, what month will it be after the moon has revolved around Earth 5 times? Use a calendar if it would be helpful.

3. How many days would be counted in total?

## Build Your Understanding

### Determine Patterns

1. Write down the next five numbers for each of these patterns.

   a) 1, 3, 7, 15, 31, ■, ■, ■, ■, ■

   b) 1, 4, 10, 22, 46, 94, ■, ■, ■, ■, ■

   c) 1, 10, 55, 280, 1405, 7030, ■, ■, ■, ■, ■

   d) 1, 6, 21, 66, 201, 606, 1821, ■, ■, ■, ■, ■

   Explain how you determined the pattern.

2. In your notebook, draw the next two possible pictures that continue the pattern below.

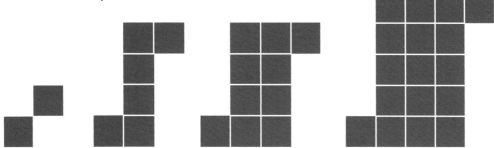

## What Did You Learn?

1. What did you do first to determine the pattern for each? Why was this your first step?

2. What other strategies can you use to find the pattern in a group of numbers?

Write the next five numbers for each of these patterns:

**1.** 2, 5, 9, 14, 20, ■, ■, ■, ■, ■

**2.** 43, 42, 45, 44, 47, 46, 49, ■, ■, ■, ■, ■

**3.** 312, 325, 339, 354, 370, ■, ■, ■, ■, ■

**4.** Create three more patterns for a classmate to figure out and continue. Be sure to complete them yourself first so you can check your classmate's answers.

Write the next number in each of these patterns:

**5.** 3, 12, 48, 192, ■

**6.** 2, 10, 50, 250, ■

**7.** 5, 15, 45, 135, ■

## Extension

Multiply these numbers. You may use a calculator if it would be helpful.

**8.** 21 x 101 = ■

**9.** 201 x 1001 = ■

**10.** 2001 x 10 001 = ■

**11.** 20 001 x 100 001 = ■

**12.** What is the pattern of these multiplication problems?

**13.** Use the same pattern to help you mentally multiply 9004 x 10 001. Check your answer with a calculator.

### Booklink

**Anno's Magic Seeds** by Mitsumasa Anno (Philomel Books: New York, NY, 1995). Jack is given two seeds: one he is supposed to eat, and one he is supposed to plant. The planted seed will produce two new seeds by the following year. But what happens when Jack decides to plant both seeds? Read this story to figure out the pattern!

# Comparing Patterns and Writing Pattern Rules

*SPACE LOG*

*PLAN:*
**You will write rules for patterns of astronaut footprints.**

*DESCRIPTION:*
In 1969, Neil Armstrong became the first human ever to set foot on the moon. As he stepped out of the Lunar Module, he said: "One small step for man, one giant leap for mankind."

There is no atmosphere on the moon, which means there is no weather. There is no wind, no rain, and no snow. The footsteps of all the astronauts who have ever walked on the moon (including Neil Armstrong's) are still there.

## Get Started

What makes this pattern a pattern?
The guideline you found is called the rule.

Discuss your rule with a classmate. Are your rules different?
How do you know which one is right?

## Compare Pattern Rules

**1.** What is the rule for this pattern?

2, 4, 6, 8, 10

**2.** What is the rule for this pattern?

3, 5, 7, 9, 11

**3.** Compare the two rules. How are they the same? How are they different?

**4.** Compare the following sets of patterns. First determine the patterns, then write the rules for each pattern. After you've completed those steps, make your comparisons.

**a)** 2, 4, 6, 8, 10  and  2, 4, 8, 16, 32

**b)** 3, 4, 3, 4, 3, 4  and  7, 9, 7, 9, 7, 9

**c)** 5, 10, 15, 20, 25  and  2, 7, 12, 17, 22

**d)** 10, 20, 30, 40, 50  and  50, 100, 150, 200, 250

## Follow Pattern Rules

**5.** Pick one number as the first number for a pattern. Make a separate number pattern for each of the three rules below. Each pattern should include five numbers.

**a)** Double the number and add 1.

**b)** Multiply each number by 2 and add 1.

**c)** Multiply each number by 3 and subtract 1.

## What Did You Learn?

**1.** What did you discover when you compared two different patterns? Will this knowledge help you find patterns more easily?

**2.** Which is easier, determining a pattern rule or following one? Why?

1. Continue the footprint pattern from Get Started by recording the next five numbers in the pattern.

2. Choose a symbol and make your own visual pattern. Include the first eight pictures of your pattern.

## Extension

3. Write a rule for the pattern shown below.

| 1 | 2 | 3 | 4 | 5 | 6 | 7 | 8 | 9 | 10 |
|---|---|---|---|---|---|---|---|---|---|
| 11 | 12 | 13 | 14 | 15 | 16 | 17 | 18 | 19 | 20 |
| 21 | 22 | 23 | 24 | 25 | 26 | 27 | 28 | 29 | 30 |
| 31 | 32 | 33 | 34 | 35 | 36 | 37 | 38 | 39 | 40 |
| 41 | 42 | 43 | 44 | 45 | 46 | 47 | 48 | 49 | 50 |
| 51 | 52 | 53 | 54 | 55 | 56 | 57 | 58 | 59 | 60 |
| 61 | 62 | 63 | 64 | 65 | 66 | 67 | 68 | 69 | 70 |
| 71 | 72 | 73 | 74 | 75 | 76 | 77 | 78 | 79 | 80 |
| 81 | 82 | 83 | 84 | 85 | 86 | 87 | 88 | 89 | 90 |
| 91 | 92 | 93 | 94 | 95 | 96 | 97 | 98 | 99 | 100 |

### Technology

Just as Neil Armstrong made footprints on the moon, you and your classmates can make footprints in sand or mud. Look at the different types of shoes worn by students in the class. What interesting patterns do you notice? Design and conduct a survey comparing these shoe soles. To prepare the survey, think about what questions you would like to find answers to. Who will you survey? How will you record the responses? What computer program might you use to present your findings? When your survey is complete, analyze the information and make some statements about your findings.

4. Create a hundreds chart like the one shown, but don't shade any of the squares. Make up more patterns of your own, and write down the rules. Give your rules to a partner along with another blank hundreds chart, and have him or her shade in the pattern.

### Booklink

**First on the Moon: What It Was Like When Man Landed on the Moon** by Barbara Hehner and Greg Ruhl (Scholastic Canada: Markham, ON, 1999). This book describes the first moon landing from the perspective of Buzz Aldrin's daughter, Jan. Buzz Aldrin was the second man to walk on the moon.

## Show What You Know

**1.** Imagine that new moon to first quarter is 7 days, first quarter to full moon is 8 days, full moon to last quarter is 7 days, and last quarter to new moon is 8 days. Calculate the pattern beginning with the first day of the month. What is the pattern for each month?

**2.** Follow this pattern:

Pick a number between 1 and 100.

Multiply by 5.

Subtract 4.

Multiply by 6.

Subtract 3.

Multiply by 7.

Subtract 2.

Explain the pattern using numbers and words.

**3.** Extend the following pattern:

# Investigating Prime Numbers

SPACE LOG

PLAN:
You will look for prime numbers among the moon's craters.

DESCRIPTION:
The moon has a lot of craters. These are formed when meteorites hit the moon. Meteorites sometimes hit Earth too, but not nearly as often because they burn up in Earth's atmosphere before they can reach the ground.

When you look at the moon from Earth, you can see some of the bigger craters. In some parts of the world, people think the moon's craters look like a face. Can you see why people talk about "the man in the moon?" Are there any other images that you see when you look at the moon?

## Get Started

1. Find the factors for these numbers:

   2   3   7   13   29   37

   What do you discover?

   The numbers above are called prime numbers. They can be divided only by 1 and themselves.

2. List the factors for these numbers. Which are prime numbers and which are composite numbers?

   10   15   3   25
   17   48   88   89

### Vocabulary

**composite number:** A whole number greater than 1 that has more than two factors. For example, 12 is a composite number, because it has six factors.
**prime number:** A whole number greater than 1 that has only two factors: 1 and itself. For example, 3 is a prime number because its only factors are 1 and itself.

### Tip

2 is the only even prime number.

### Find Prime Numbers

Look at this illustration.

1. Find the shortest path from start to finish, stepping only on the craters that have prime numbers.

2. Record the path you took.

3. Compare your path with a classmate's.

1. Why is 2 the only even prime number?

2. Explain the relationship between factors and prime numbers.

3. In your notebook, write some examples of composite numbers. Explain the difference between composite numbers, factors, and prime numbers.

## Practice

**1.** Make a chart in your notebook like the one below.

| Prime | Composite |
|-------|-----------|
|       |           |

**2.** Put each of the following numbers into the correct column.

5, 83, 47, 2, 69, 90, 4, 18, 23, 21

**3.** Copy these sets of numbers into your notebook, and circle the prime numbers in each set.

**a)** 49, 50, 51, 52, 53, 54

**b)** 63, 64, 65, 66, 67, 68

**c)** 95, 96, 97, 98, 99

**d)** 69, 70, 71, 72, 73, 74, 75, 76

**e)** 36, 37, 38, 39, 40, 41

**f)** 77, 78, 79, 80, 81, 82, 83, 84

### Technology

You have learned many more terms, concepts, and skills since you began your personal reference guide back in Chapter 1. Add to your document, remembering to use the same format as when you started.

**4.** Copy and complete a chart like the one below into your notebook to describe prime numbers, factors, and composite numbers. List the characteristics of each concept, and include an example.

| Prime Numbers | Factors | Composite Numbers |
|---------------|---------|-------------------|
|               |         |                   |

**5.** Make a chart with your class including everyone's responses. The chart can be put up somewhere in the classroom for future reference.

## Lesson 5

# Exploring Multiples and Factors

SPACE LOG

PLAN:
You will practise finding multiples and factors as you study weight on the moon.

DESCRIPTION:
The astronauts who went to the moon experienced gravity that is $\frac{1}{6}$ of Earth's gravity. This means that while they were on the moon they weighed only $\frac{1}{6}$ of what they weighed here on Earth. It also meant they could travel 6 times higher and farther with the same amount of effort as they would use on Earth.

## Get Started

Phillip and Ruth had a lot of fun at the exhibit about the moon's gravity. The display included a mini-trampoline that gave students who jumped on it an idea of what gravity on the moon is like.

1. If you can jump 1 m high on Earth, how far could you jump if you were on the moon?

2. If you weighed 30 kg on the moon, how much would you weigh on Earth?

3. Which of these numbers are multiples of 6?

   12   51   99   120   43   48

4. List all the numbers that divide into 6 with no remainder. These numbers are called factors.

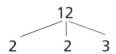
## Build Your Understanding

## Factor Whole Numbers

Here's a factor tree for 12. The numbers at the ends of the "branches" will multiply together to equal 12.

```
      12
    / | \
   2  2  3
```

The tree shows you that the factors of 12 are $2 \times 2 \times 3$.

**1.** Complete these factor trees:

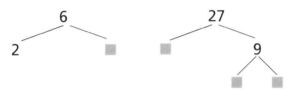

**2.** Draw factor trees for these numbers:

**a)** 40 **b)** 50 **c)** 44 **d)** 28 **e)** 14 **f)** 35

### Technology

Create electronic factor trees using a draw computer program, graphic organizer, or desktop publisher. Copy and paste the factor lines as needed. Use the "group objects" tool to lock your factor trees. Then try joining your factor trees to make larger factor trees.

### Vocabulary

**prime factor:** A number that cannot be factored any further than itself and 1. Examples of prime factors are 2, 3, 5, 7, 11.

**prime number:** A whole number that has only two factors: 1 and itself

### Journal

The number 1 is a factor of any number but it is not shown in factor trees. Explain why.

## What Did You Learn?

**1.** Is 2 a factor of 9? Explain your answer using numbers, pictures, and words.

**2.** Can you find a number that is a multiple of 100, but not of 10? Explain your answer using numbers, pictures, and words.

**3.** Explain how you know when a factor tree is finished.

**4.** Explain how you figured out your weight on Earth based on your weight on the moon.

## Practice

**1.** Find the prime factors of these numbers:

6   12   18   24   30

What do they have in common?

**2.** What do these numbers have in common?

16   24   40   88   800

**3.** What do these numbers have in common?

10   30   70   20   40

**4.** What do these numbers have in common?

28   42   70   98   126

In your notebook, list the factors of these numbers:

**5.** 64   **6.** 72   **7.** 50   **8.** 66   **9.** 56

**10.** 48   **11.** 60   **12.** 78   **13.** 96   **14.** 100

**15.** Think of three numbers that only have two factors. Share your answers with a classmate.

### Math Problems to Solve

**16.** The product is 36. What are the factors?

**17.** Two of the factors are 2 and 4. If there are three factors, and the third factor is a number between the first two factors, what is the product?

**18.** There are four factors with a product of 48. What are the possible combinations of factors that multiply to 48?

**Lesson 6**

# Finding Missing Terms

*SPACE LOG*

*PLAN:*
You will learn about missing terms and the relationship between multiplication and division. You will practise solving problems about golf on the moon.

*DESCRIPTION:*
Since the moon's gravity is only $\frac{1}{6}$ that of Earth's, playing a game like golf could be a lot of fun! Alan Shepard hit two golf balls on the moon in 1971.

## Get Started

Copy these problems into your notebook and fill in the blanks.

**1.** $1 + \blacksquare = 4$     **2.** $5 + \blacksquare = 10$

**3.** $4 - \blacksquare = 1$     **4.** $8 - \blacksquare = 7$

**5.** $2 \times \blacksquare = 6$     **6.** $3 \times \blacksquare = 12$

**7.** $\blacksquare \times 10 = 10$     **8.** $12 \times \blacksquare = 36$

**9.** $12 \div \blacksquare = 2$     **10.** $\blacksquare \div 5 = 7$

**11.** $96 - \blacksquare = 36$     **12.** $\blacksquare \times 4 = 64$

**13.** $\blacksquare + 17 = 35$     **14.** $54 - \blacksquare = 27$

**15.** $4 \times \blacksquare = 44$     **16.** $\blacksquare - 11 = 39$

**Journal**

Explain how you know which numbers belong in the blanks.

## Build Your Understanding

### Solve Equations

Li and Eric decided they would play an imaginary game of moon golf. If Li can hit a golf ball 200 m on Earth, how far would it go on the moon? If Eric hit his golf ball 900 m on the moon, how far could he have hit it on Earth?

Li wrote her problem like this:

200 m x 6 = ◼ m

To find her solution, she multiplied 200 by 6. On the moon, she can hit a golf ball 1200 m.

Eric wrote his problem like this:

900 m ÷ 6 = ◼ m

Eric solved his problem by dividing 900 by 6. On Earth, he can hit a golf ball 150 m.

Li's problem rewritten as a division equation would look like this:

1200 m ÷ 6 = 200 m

In your notebook, rewrite Eric's problem as a multiplication equation.

900 m ÷ 6 = 150 m

Remember that gravity on the moon is $\frac{1}{6}$ of Earth's gravity.

1. Solve these equations:

   a) 800 ÷ 4 = ◼

   b) 600 ÷ 30 = ◼

   c) 550 ÷ ◼ = 22

   d) ◼ ÷ 12 = 40

2. Now show each division equation above as a multiplication equation.

3. Solve these equations:

   a) 30 x 5 = ◼

   b) 62 x 3 = ◼

   c) 800 x ◼ = 2400

   d) ◼ x 40 = 2400

4. Now show each multiplication equation above as a division equation.

## What Did You Learn?

**1.** What is the relationship between a multiplication problem and a division problem?

**2.** What do these equations all have in common?

**a)** 10 x 6 = ▦    **b)** ▦ ÷ 6 = 10    **c)** ▦ ÷ 10 = 6

**3.** Write a list of steps to follow when calculating distance on Earth compared to distance on the moon.

## Practice

Find the missing term for each of the questions below:

**1.** 1 + ▦ = 8          **2.** ▦ + 5 = 7          **3.** 10 − ▦ = 5

**4.** ▦ − 14 = 2         **5.** 4 x ▦ = 12         **6.** 12 x ▦ = 24

**7.** 12 ÷ ▦ = 1         **8.** 45 ÷ ▦ = 9         **9.** 30 ÷ ▦ = 6

**10.** 32 x ▦ = 64       **11.** 30 ÷ 3 = ▦        **12.** 256 − 48 = ▦

**13.** 2 x ▦ = 82        **14.** 63 ÷ 9 = ▦        **15.** 14 ÷ ▦ = 7

**16.** ▦ x 5 = 25        **17.** 27 x ▦ = 81       **18.** ▦ x 6 = 54

### Extension

Imagine that you are doing a study comparing gravity on the moon to gravity here on Earth.

**19.** Choose an activity that you enjoy doing, such as throwing a baseball or football, hitting a tennis ball, high jump, long jump, or running.

**20.** Find a classmate who has chosen the same activity that you have, and work together to measure the distance that each of you achieves in performing tasks for the chosen activity. For example, if you chose baseball as an activity that you enjoy doing, you might measure how far you can throw a baseball, how far you can run in 10 seconds, or how far you can hit a baseball. Find measurements for at least three tasks related to your activity. Record your results on a piece of paper.

**21.** Now figure out what your results would be on the moon.

**22.** Share your results with the class. Use a computer program to create a chart or graph to show your results.

## Show What You Know

**Review: Lessons 4 to 6, Problem Solving, Multiples, Factors, Prime Numbers, and Composite Numbers**

**1. a)** Gravity on planet X is $\frac{1}{8}$ of gravity on Earth. If you weighed 50 kg on Earth, what would your weight be on planet X?

   **b)** If you can hit a golf ball on Earth 168 m, how far can you hit a golf ball on planet X?

**2.** In the Extension in Lesson 6, you calculated the distance achieved on Earth when performing a certain activity like throwing a baseball or hitting a golf ball, and also the distance achieved when performing this activity on the moon. Now calculate the distance achieved when performing the activity you chose on planet X.

**3.** In your own words, explain

   **a)** multiples

   **b)** factors

   **c)** prime numbers

   **d)** composite numbers

   Give two examples of each.

**Lesson 7**

# Converting Fractions to Decimals and Percents

SPACE LOG

PLAN:
You will convert fractions to decimals and percents while calculating the distance to Mars.

DESCRIPTION:
NASA is planning a mission to Mars! It will take between 4 and 6 months to get there. Then the crew will spend 18 months on Mars, exploring and doing experiments. It will take the crew another 4 to 6 months to return to Earth. By the time they return, the crew will have spent over 2 years in space!

## Get Started

**1.** Write a fraction for each of the sections shown below.

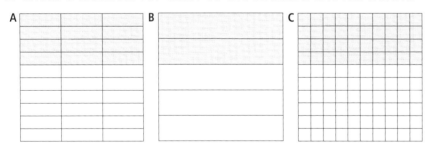

A   B   C

**2.** What do all of these fractions have in common?

**3.** Can you think of a rule for reducing each of the fractions to the same number?

## Build Your Understanding

### Convert Fractions to Percents

Fractions are often written with a denominator of 100. This means the fractions have a value out of 100, or per hundred. "Percent" is an abbreviation for the Latin words "per centum" meaning "per hundred."

You have seen the word "cent" in English words before. Century, centipede, and cents are some examples. What do these words have in common?

**Vocabulary**

**equivalent fractions:** Fractions that name the same amount or part. For example, $\frac{2}{4}$, $\frac{4}{8}$, and $\frac{3}{6}$ are all equivalent fractions; they can be reduced to $\frac{1}{2}$.

**percent:** A number out of 100. It is written with the percent sign (%). For example, 18 percent is $\frac{18}{100}$ and is written as 18%.

**1.** Complete these equations:

a) $\frac{4}{5} = \frac{\blacksquare}{100}$   b) $\frac{2}{10} = \frac{\blacksquare}{100}$   c) $\frac{3}{4} = \frac{\blacksquare}{100}$   d) $\frac{1}{4} = \frac{\blacksquare}{100}$

e) $\frac{8}{10} = \frac{\blacksquare}{100}$   f) $\frac{1}{2} = \frac{\blacksquare}{100}$   g) $\frac{4}{25} = \frac{\blacksquare}{100}$   h) $\frac{2}{5} = \frac{\blacksquare}{100}$

Fractions can be converted to decimals:

$\frac{45}{100} = 0.45$

**2.** Convert each of your solutions in question 1 to a decimal.

Decimals can be converted to percents:

$0.45 = 45\%$

The fraction $\frac{4}{5}$ can be written as $\frac{80}{100}$. This can be said as 80 out of 100 or 80 per hundred. This is the same as 80 percent or 80%.

Another way to write $\frac{4}{5}$ as a percent would be just to change it to a decimal number.

$\frac{4}{5} = \frac{80}{100} = 0.80$. This is the same as 80%.

**3.** Write the remaining fractions in question 1 as percents.

**4.** Write each of the following as a fraction, a decimal, and a percent. The first one has been done for you.

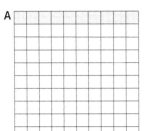 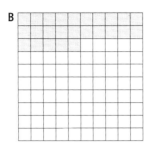 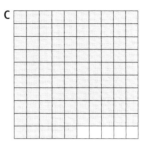

A) $\frac{10}{100} = 0.10 = 10\%$

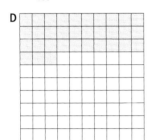 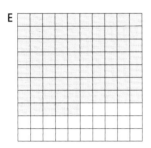

## What Did You Learn?

**1.** Explain the steps used to write a fraction such as $\frac{3}{5}$ as a percent.

**2.** Explain the steps used to write a decimal number such as 0.25 as a percent.

**3.** Write an equation for this problem:

If a mission to Mars takes 30 months, and the crew spends 6 months getting to Mars and 6 months getting back, what percent of the mission is spent travelling?

Now solve the problem.

**4.** In which form do you find numbers easiest to understand, fractions, decimals, or percents? Explain your answer.

Find the missing numerator:

**1.** $\frac{3}{5} = \frac{\blacksquare}{100}$   **2.** $\frac{25}{100} = \frac{\blacksquare}{4}$   **3.** $\frac{20}{100} = \frac{\blacksquare}{10}$   **4.** $\frac{5}{10} = \frac{\blacksquare}{100}$

**5.** $\frac{10}{100} = \frac{\blacksquare}{20}$   **6.** $\frac{5}{100} = \frac{\blacksquare}{20}$   **7.** $\frac{75}{100} = \frac{\blacksquare}{4}$   **8.** $\frac{80}{100} = \frac{\blacksquare}{25}$

Convert these fractions to decimals:

**9.** $\frac{30}{100}$   **10.** $\frac{7}{8}$   **11.** $\frac{6}{25}$   **12.** $\frac{2}{4}$   **13.** $\frac{4}{5}$

Convert these fractions to decimals and then to percents:

**14.** $\frac{2}{5}$   **15.** $\frac{5}{10}$   **16.** $\frac{45}{100}$   **17.** $\frac{82}{100}$   **18.** $\frac{1}{10}$

Reduce these fractions to their lowest terms:

**19.** $\frac{3}{6}$   **20.** $\frac{25}{100}$   **21.** $\frac{2}{4}$   **22.** $\frac{6}{12}$   **23.** $\frac{5}{25}$

## A Math Problem to Solve

**24. a)** Phillip has bought a bag of 100 "space snacks" at the Science Centre souvenir shop. He wants to share the snacks with 3 classmates. If he gives Claire $\frac{2}{5}$, Ajay $\frac{2}{10}$, and Li $\frac{1}{4}$, how many snacks will be left for him? What percent of the snacks did each person get? Show your work.

**b)** List the steps you followed to figure out this problem.

## Lesson 8

# Converting Decimals to Percents

*SPACE LOG*

*PLAN:*
You will convert fractions and decimals to percents, and percents to decimals and fractions, while learning about the atmosphere on Mars.

*DESCRIPTION:*
The atmosphere of Mars is almost all carbon dioxide. Since humans need oxygen to breathe, any mission to that planet would have to bring along its own supply of oxygen, or make oxygen from the carbon dioxide in Mars' atmosphere.

## Get Started

**1.** Convert these fractions to decimals:

**a)** $\dfrac{25}{100}$   **b)** $\dfrac{4}{10}$   **c)** $\dfrac{8}{10}$   **d)** $\dfrac{33}{100}$

**e)** $\dfrac{1}{100}$   **f)** $\dfrac{2}{10}$   **g)** $\dfrac{45}{100}$   **h)** $\dfrac{99}{100}$

**2.** Convert these decimals to percents. The first one is done for you.

**a)** 0.25 = 25%  **b)** 0.1   **c)** 0.5   **d)** 0.89

**e)** 0.33   **f)** 0.04   **g)** 0.75   **h)** 0.56

### Journal

Describe in your own words how fractions, decimals, and percents are all related.

## Convert Percents and Decimals to Fractions

Here's a table showing the composition of Mars' atmosphere:

| Carbon Dioxide | 95% |
|---|---|
| Nitrogen | 3% |
| Argon | 2% |

1. Write each of these amounts as a fraction out of 100. Reduce each to lowest terms.

2. Write each of the amounts as a decimal.

## Convert Decimals to Percents

3. To compare Mars' atmospheric composition to Earth's, here is a table showing the composition of Earth's atmosphere. The data for "Other Gases" is missing. Can you figure out what it might be, based on the other percentages in the chart?

| Nitrogen | 78% |
|---|---|
| Oxygen | 21% |
| Other Gases | |

4. Convert these percents to fractions:

   **a)** 36%    **b)** 89%    **c)** 5%

5. Convert these percents to decimals:

   **a)** 56%    **b)** 2%    **c)** 99%

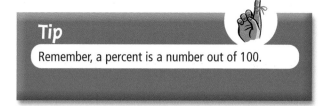

**Tip**

Remember, a percent is a number out of 100.

## What Did You Learn?

1. Explain how you figured out the percent of other gases in the composition of Earth's atmosphere.

2. Explain in your own words how to convert a fraction to a decimal.

3. Explain in your own words how to convert a decimal to a percent.

4. Explain in your own words how to convert a percent to a fraction and a decimal.

Convert these fractions to percents:

**1.** $\frac{25}{100}$    **2.** $\frac{1}{2}$    **3.** $\frac{3}{4}$    **4.** $\frac{1}{4}$

**5.** $\frac{4}{25}$    **6.** $\frac{10}{25}$    **7.** $\frac{1}{8}$    **8.** $\frac{8}{10}$

**9.** Estimate how much of each square is shaded. Show your estimates as fractions, decimals, or percents.

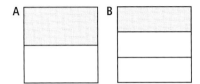
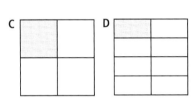

**Technology**

Work with a classmate to gather a collection of 100 items together into a group. These might be books, supplies, pencils and pens, or other items that you find. Now sort these items in various ways, for example, by colour, type, or shape. Use a desktop publisher or draw computer program to show the different ways you sorted the items. Convert your data to fractions, decimals, and percents, and record your data in a chart.

Convert these percents to fractions and then to decimals:

**10.** 56%    **11.** 23%    **12.** 12%    **13.** 9%

**14.** 98%    **15.** 82%    **16.** 74%    **17.** 43%

**18.** 28% of a pizza is covered in onions and mushrooms. 45% of the pizza is covered in tomatoes and pepperoni. The remainder of the pizza is covered in green peppers and sausage. What percentage is covered in green peppers and sausage?

## Extension

**19.** Ajay's score on his math test was $\frac{45}{50}$. What is his score as a percent?

**20.** Show Ajay's score as a decimal.

**21.** What percentage of the test did Ajay have difficulty with?

**22.** List the steps you followed to solve this problem.

**23.** Imagine that in Ajay's school the marking scheme shown in the chart is used. What word describes Ajay's score on this math test?

| 60% – 69% | Satisfactory |
| --- | --- |
| 70% – 79% | Good |
| 80% – 89% | Very Good |
| 90% – 100% | Excellent |

Lesson 8: Converting Decimals to Percents

## Lesson 9

# Multiplying by 0.1, 0.01, and 0.001

SPACE LOG

PLAN:
You will multiply numbers by 0.1, 0.01, and 0.001 mentally.

DESCRIPTION:
Every mission to space — whether it's spending time on the International Space Station, orbiting Earth in the space shuttle, or planning a long-term mission to Mars — requires math skills. Astronauts use math to navigate their spacecraft, to calculate how many supplies they need to take along on the mission, and to do their experiments. The Grade 6 students learn these things as they continue to walk through the space exhibit at the Science Centre.

**Canadian astronaut Chris Hadfield**

## Get Started

Ajay and Mahalia sat down on a bench in the space exhibit. They were exhausted, and they hadn't even seen all the exhibits yet.

"I had no idea how much math was involved in exploring space," said Mahalia.

"I know!" exclaimed Ajay. "It's incredible!"

Multiplication can be used a lot when using math in space. As you have learned, multiplication means groups of something.

Multiplying decimals means finding a fraction or part of a whole number.

0.1 x 40 means one tenth of 40.

0.01 x 40 means one hundredth of 40.

0.001 x 40 means one thousandth of 40.

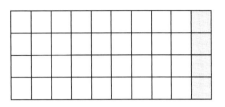

**1.** Look at these equations:

- 1000 x 4 = 4000
- 100 x 4 = 400
- 10 x 4 = 40

**a)** What pattern do you notice?

**b)** What rule can you state about multiplying by 10, 100, and 1000?

**c)** Share your rule with a classmate.

**2.** Now look at these equations:

- 0.1 x 4 = 0.4
- 0.01 x 4 = 0.04
- 0.001 x 4 = 0.004

**a)** What pattern do you notice?

**b)** What rule can you state about multiplying by 0.1, 0.01, and 0.001?

**c)** Share your rule with a classmate.

## Build Your Understanding

## Multiply Whole Numbers by Tenths, Hundredths, and Thousandths

**1.** Answer these questions mentally:

**a)** 0.1 x 2 =   **b)** 0.1 x 5 = ▧  **c)** 0.01 x 6 = ▧

**d)** 0.001 x 8 = ▧  **e)** 0.01 x 45 = ▧  **f)** 0.1 x 456 = ▧

**g)** 0.001 x 879 = ▧  **h)** 0.1 x 78 = ▧  **i)** 0.001 x 3456 = ▧

**2.** If 162 is multiplied by 0.01, what digit will be in the tenths place?

**3.** If 187 is multiplied by 0.001, what digit will be in the thousandths place?

**4.** Copy a place-value chart like this one into your notebook.

| Hundreds | Tens | Ones | Tenths | Hundredths | Thousandths |
|----------|------|------|--------|------------|-------------|
|          |      |      |        |            |             |

Answer these questions mentally and show each product on your place-value chart.

a) 0.001 x 456 = ▨    b) 0.01 x 34 = ▨    c) 0.1 x 90 = ▨

d) 0.01 x 56 = ▨    e) 0.001 x 3 = ▨    f) 0.01 x 5122 = ▨

g) 0.1 x 3459 = ▨

## What Did You Learn?

1. List the steps you followed to mentally multiply numbers by 0.1, 0.01, and 0.001.

2. How did you come up with a rule to help you? How is your rule like a pattern rule?

3. In your notebook, draw a picture to show one tenth of 30. Now draw a picture the same size as your first picture to show one hundredth of 300. What do you notice?

## Practice

Answer these questions mentally:

1. 0.1 x 3 = ▨    2. 0.01 x 3 = ▨    3. 0.001 x 3 = ▨

4. 0.1 x 45 = ▨    5. 0.01 x 456 = ▨    6. 0.001 x 7898 = ▨

7. 0.1 x 671 = ▨    8. 0.01 x 5123 = ▨

9. a) If 1678 is multiplied by 0.01, what digit will be in the tenths place?

   b) What digit will be in the hundredths place?

10. a) If 782 is multiplied by 0.001, what digit will be in the tenths place?

    b) What digit will be in the hundredths place?

    c) What digit will be in the thousandths place?

11. a) Multiply these measurements by 0.1, 0.01, and 0.001.

       3476 cm      2357 m      362 mm

    b) What patterns do you notice?

    c) What measurement units apply to the new numbers?

    d) Show your data in a chart, and share with a classmate.

**Lesson 10**

# Dividing Decimals by Tens and Hundreds

*SPACE LOG*

*PLAN:*

You will use the data about patterns and about Sojourner's length to divide by tens and hundreds.

*DESCRIPTION:*

On July 4, 1997, a remote space probe called Pathfinder landed on Mars to explore. It unfolded its panels and inside was a little robot named Sojourner. This little robot explored the planet and sent photographic images back to Earth. Sojourner was only 0.65 m long, 0.48 m wide, and 0.3 m high. It weighed 10 kg.

## Get Started

1. *Sojourner* was 0.65 m long. The average car is about 4.5 m long. Approximately how many Sojourners, laid end to end, would equal the length of a car?

2. Can you think of another way to figure this out?

**Journal**

How is it helpful to know the size of *Sojourner* in relationship to something you see every day?

3. Use a calculator to see how close your approximation is.

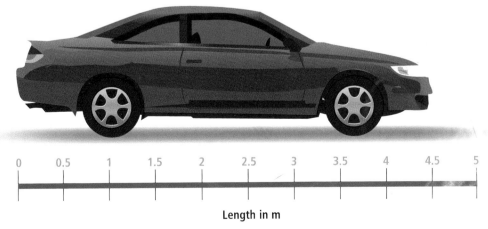

0    0.5    1    1.5    2    2.5    3    3.5    4    4.5    5

**Length in m**

### Divide Decimals

1. Solve these questions mentally. Some of them have been completed for you.

   **a)** 0.8 ÷ 1 = ■

   **b)** 0.8 ÷ 10 = 0.08

   **c)** 0.8 ÷ 100 = 0.008

2. Do you see a pattern? What is it?

3. Now solve these questions:

   **a)** 0.5 ÷ 1 = ■

   **b)** 0.5 ÷ 10 = ■

   **c)** 0.5 ÷ 100 = ■

4. Do you see a pattern? What is it?

### Journal

In your journal, write down the patterns you noticed. What rule can you state about dividing a decimal number by 10 and 100?

## What Did You Learn?

1. How does knowing the pattern help you solve these problems more easily?

2. Write a rule for the pattern of the division problems in Build Your Understanding.

3. Explain how you divided a decimal number by 10 and 100.

### Technology

Add the patterns and rules that you have learned to your personal reference guide. Don't forget to save the updated version.

## Practice

Solve these problems:

**1.** $0.234 \div 10 =$ ▨

**2.** $32.356 \div 100 =$ ▨

**3.** $351.023 \div 100 =$ ▨

**4.** $4.310 \div 10 =$ ▨

**5.** $653.25 \div 10 =$ ▨

**6.** $874.603 \div 100 =$ ▨

**7.** $34.272 \div 10 =$ ▨

**8.** $10.894 \div 10 =$ ▨

**9.** $6.534 \div 100 =$ ▨

**10.** $5356.054 \div 100 =$ ▨

**11.** Explain the pattern that can help you solve these problems.

**12.** Remember that *Sojourner* was 0.65 m long. Pick an object in your classroom or at home, such as a desk, couch, or chair, and measure its length. How does this length compare to *Sojourner*'s?

**13.** Write three math questions comparing *Sojourner* to the object you measured, and give them to a classmate to answer.

## Extension

Dividing by decimals means dividing a whole number by part of a whole number. Since you are dividing by less than a whole number, you will have more parts, or a greater number in your answer, than you started with.

$2 \div 0.1$ means 2 divided by one tenth.

If you have 2 pieces of string, how many one-tenth pieces can you cut? We know that each piece must be divided into 10 parts.
2 pieces of string, each with 10 parts, is the same as saying 2 x 10.

$2 \times 10 = 20$

Therefore, $2 \div 0.1 = 20$

If you have 4 cakes, how many one-hundredth pieces can you cut? We know that each cake must be divided into 100 parts.
4 cakes, each with 100 parts, is the same as saying 4 x 100.

$4 \times 100 = 400$

Therefore, $4 \div 0.01 = 400$

**14.** Look at these equations:

- $4 \div 0.1 = 40$
- $4 \div 0.01 = 400$
- $4 \div 0.001 = 4000$

**a)** What pattern do you notice?

**b)** What rule can you state about dividing by 0.1, 0.01, and 0.001?

**c)** Share your rule with a classmate.

**15.** Copy a place-value chart like this one into your notebook.

| Hundred Thousands | Ten Thousands | Thousands | Hundreds | Tens | Ones |
|---|---|---|---|---|---|
|  |  |  |  |  |  |

Answer these questions mentally and show each product on your place-value chart.

**a)** $456 \div 0.1 = $ ■   **b)** $42 \div 0.01 = $ ■

**c)** $234 \div 0.001 = $ ■   **d)** $12 \div 0.1 = $ ■

**e)** $2698 \div 0.1 = $ ■   **f)** $90 \div 0.001 = $ ■

## Show What You Know

### Review: Lessons 7 to 10, Fractions, Decimals, and Percents

**1. a)** Phillip bought a bag of 50 space snacks. If he shared $\frac{1}{5}$ of the snacks with Claire, $\frac{3}{10}$ with Ajay, and $\frac{8}{25}$ with Li, how many snacks will be left for him?

**b)** List the steps you followed to solve 1a).

**2.** Ahanu scored $\frac{18}{25}$ on his Science test. Show Ahanu's mark as a decimal and as a percentage. Use the marking scheme in Lesson 8 to determine the comment that Ahanu received for his mark.

**3.** *Sojourner* is 0.65 m long and 0.3 m high. Now imagine that a car is 3.5 m long and 1.5 m high. Using Sojourner's length measurement, calculate how many Sojourners would fit into the length of a car. Now use *Sojourner's* height measurement to calculate how many Sojourners would fit into the height of a car.

# Chapter Review

**Chapter 2**

1. Determine the pattern rule and write the next five numbers in each pattern.

   **a)** 2, 5, 11, 23,

   **b)** 2, 6, 12, 20, 30

   **c)** 3, 3, 6, 9, 15, 24

2. Explain the rule for this pattern:

   1, 3, 7, 15, 31

3. Write the next five numbers in this pattern:

   3, 6, 10, 15, 21

4. In your notebook, draw the next two pictures to continue this pattern.

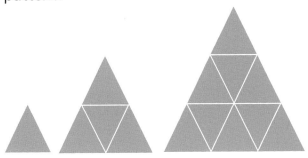

5. List all the prime numbers to 100.

6. List all the composite numbers to 100.

7. List all the factors of 48.

8. Which of these numbers are multiples of 12?

   156, 26, 120, 48, 6, 144

9. Draw factor trees for each of these numbers:

   **a)** 24   **b)** 64   **c)** 16   **d)** 39   **e)** 34   **f)** 99

**10.** Find the missing term:

a) 16 + ■ = 32    b) 95 − ■ = 5

c) 90 x ■ = 270    d) 99 ÷ ■ = 33

e) 99 + ■ = 100    f) 75 + ■ = 100

g) 48 − ■ = 40    h) 105 ÷ ■ = 5

i) 450 ÷ ■ = 9    j) 19 − ■ = 10

k) 8 x ■ = 128    l) 5 x ■ = 125

**11.** Convert the following fractions to decimals and percents:

a) $\frac{25}{100}$    b) $\frac{9}{10}$    c) $\frac{4}{5}$    d) $\frac{20}{25}$    e) $\frac{4}{10}$

**12.** Convert the following decimals to percents:

a) 0.99%    b) 0.51    c) 0.75    d) 0.48    e) 0.16

**13.** Answer these multiplication questions mentally:

a)    4    b)    9    c)    42    d)    2

x 0.2    x 0.1    x 0.01    x 0.3

e) 0.001 x 3 = ■    f) 0.1 x 7925 = ■

g) 0.01 x 491 = ■    h) 0.001 x 863 = ■

**14.** Solve these division questions mentally:

a) 0.629 ÷ 10 = ■    b) 50.377 ÷ 100 = ■    c) 62 499.3 ÷ 100 = ■

d) 5.178 ÷ 10 = ■    e) 952.15 ÷ 10 = ■    f) 315.420 ÷ 100 = ■

Chapter
**2**

# Chapter Wrap-Up ○ △ ▢ ⬠

You have reached the end of Chapter 2. In this chapter, you looked at patterns and determined their rules, and you learned how to factor, how to solve for a missing term, and how to find prime numbers. You also found out how decimals and percents are related, and how multiplying and dividing decimals and whole numbers are related.

For this project, you are going to design a game that uses all the different types of math problems that you learned in this chapter. You are going to use the solar system as a theme for the game.

## Plan Your Game

1. Work in small groups.

2. Decide on a design for your game. Will it be a board game or a game that can be played outdoors? How will you decide what the game rules will be?

3. Review all of the math concepts from this chapter. What did you learn? Questions used in your game should cover all of these concepts.

4. Who will write the math problems? Who will check that the answers are correct?

## Play Your Game

1. When you are finished, invite another group to play your game.

2. Are there any ways that your group can improve your game? If yes, how?

Good luck!

# Chapter 3

# Math Beyond Mars

To explore space beyond Mars, we continue to rely on our math skills. Math helps us understand what we see through telescopes, or what we discover through probes sent into deep space.

In this chapter, you will

- represent large numbers
- practise solving for missing values
- write and solve simple equations
- collect, interpret, and compare data
- calculate mean, median, and mode

At the end of this chapter, you will use the math skills you have learned to plan a research project about a planet or star of your choice.

Answer these questions to prepare for your adventure:

1. On a place-value chart, which three columns come after the hundred-thousands column? Draw a place-value chart to help you.

2. Why do you think a missing term is sometimes represented by a letter? For example, $36 \div n = 12$.

3. In your own words, explain mean, median, and mode.

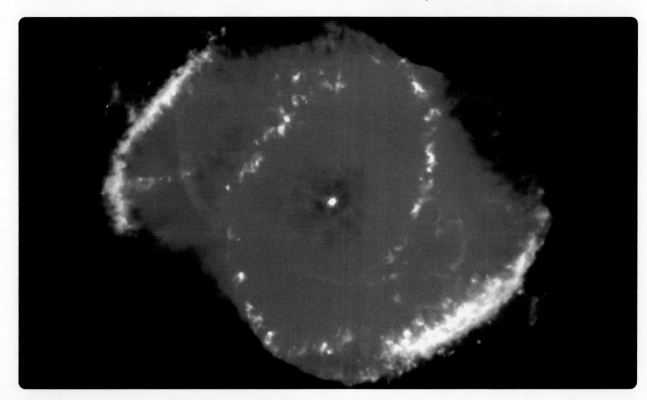

## Lesson 1

# Whole Numbers to 1 000 000

SPACE LOG

PLAN:
You will represent and compare whole numbers to 1 000 000 while studying the sizes of different planets.

DESCRIPTION:
Venus and Earth are almost the same size. Mercury and Mars are also almost the same size. Along with Pluto, these four planets are the smallest in the solar system.

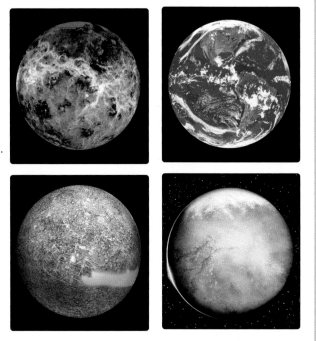

## Get Started

**1.** Write each of the following in numeral form:
   **a)** eight hundred seventy-five thousand fourteen
   **b)** fifty-four thousand nine hundred sixty-two
   **c)** four thousand five hundred two
   **d)** seventeen thousand two hundred one
   **e)** one hundred ninety-nine thousand two hundred ninety-one
   **f)** one hundred forty-three thousand two hundred four

**2.** Write these numbers in words:
   **a)** 465 007      **b)** 6445
   **c)** 980 456      **d)** 57 404
   **e)** 96 331       **f)** 4579

## Build Your Understanding

### Represent Large Numbers

This table shows the diameters of the four terrestrial planets:

| Planet | Diameter (in km) |
|---|---|
| Mercury | 4 880 |
| Venus | 12 100 |
| Earth | 12 756 |
| Mars | 6 794 |

**1.** Put the planets in order from smallest to largest.

**2.** Write out each number in expanded form.

**3.** Write each number on a place-value chart.

**Vocabulary**

**expanded form:** A number written to show the value of each digit; for example, 351 246 = 300 000 + 50 000 + 1000 + 200 + 40 + 6

**standard form:** A number written as a numeral; for example, 351 246

## What Did You Learn?

1. Which planet's diameter length will fit at least twice into two others' diameter lengths?

2. Which planet has a diameter more than half the size of two of the other planets?

3. Explain the steps you used to answer questions 1 and 2.

4. Compare your list of steps with a classmate's. How are they the same, and how are they different?

5. Explain standard form and expanded form in your own words.

## Practice

Write these numbers in words:

**1.** 34 789 **2.** 3412 **3.** 99 898

**4.** 432 767 **5.** 100 000

Write these numbers in expanded form:

**6.** 12 356 **7.** 678 756 **8.** 98 213

Write the following numbers as numerals:

**9.** thirty thousand

**10.** five thousand six hundred eighty-two

**11.** ninety-nine thousand nine hundred ninety-nine

### Extension

Write down the following numbers:

**12.** The number that is 10 000 less than 87 572

**13.** The number that is 1000 greater than 198 774

**14.** The number that is 100 000 less than 312 786

**15.** The number that is 10 greater than 64 546

**16.** The number that is 10 000 less than 102 337

**17.** The number that is 100 000 greater than 334

Lesson 2

# Whole Numbers Greater Than 1 000 000

SPACE LOG

PLAN:
You will represent and compare whole numbers greater than 1 000 000 while studying the distance between planets and the sun, and other space data.

DESCRIPTION:
Mercury, Venus, Earth, and Mars are called the "terrestrial" planets because they are all made primarily of rock.

Mount Rundle in Banff, Alberta, is part of the Rocky Mountain range. Mercury, Venus, Earth, and Mars are all rocky planets.

## Get Started

At the Science Centre, Phillip and Ruth decided they needed a way to organize all the information they were learning about the planets Mercury, Venus, Mars, and Earth. They took out their notebooks and began taking notes.

One of the exhibits they saw showed the size of the planets. They found that the distances between these planets and the sun are represented by very large numbers.

Copy and complete the place-value chart below for each planet.
The numbers represent how far each planet is from the sun.

| Planet | Km | Hundred Millions | Ten Millions | Millions | Hundred Thousands | Ten Thousands | Thousands | Hundreds | Tens | Ones |
|--------|-----|-----------------|--------------|----------|------------------|---------------|-----------|----------|------|------|
| Mercury | 58 000 000 | | | | | | | | | |
| Venus | 108 000 000 | | | | | | | | | |
| Earth | 150 000 000 | | | | | | | | | |
| Mars | 228 000 000 | | | | | | | | | |

## Build Your Understanding

### Read Large Numbers

The distances used in Get Started are rounded off. This means they all end in 0. Astronauts often have to work with numbers that aren't rounded off. Answer the following questions using these numbers. First, take turns reading the numbers aloud with a classmate.

### Vocabulary

**rounding:** A rule used to make a number approximate to a certain place value. Numbers are rounded up when the digit immediately following the required place value is 5 or higher, and rounded down when the digit is less than 5. For example, 238 rounded to the nearest ten is 240. 1238 rounded to the nearest hundred is 1200.

| 6 518 984 | 108 011 | 4 587 904 | 1 807 654 | 1 024 556 |
|-----------|---------|-----------|-----------|-----------|
| 5 065 445 | 12 756 | 2 556 009 | 259 189 | 5 843 880 |

1. Write down the numbers in order from least to greatest.

2. Make a place-value chart. Record each number in the place-value chart.

3. Which numbers are greater than (>) 999 999?

4. Which numbers are less than (<) 999 999?

5. Which numbers are > 99 999 but < 999 999?

6. Which number has an 8 in the thousands place and a 1 in the tens place?

7. Which number has a 5 in the hundred thousands place, a 6 in the thousands place, and a 9 in the ones place?

8. Round these numbers to the nearest
   a) ten: 85, 71, 102
   b) hundred: 632, 596, 1213
   c) thousand: 9932, 8431, 12 813

## What Did You Learn?

1. What patterns did you notice in your place-value chart? Explain.

2. What did you find easy about reading the numbers in this lesson's Build Your Understanding, and what did you find most challenging? Explain your answers.

## Practice

1. Use numbers from 1 to 6 to make up ten different large numbers greater than 1 000 000. Arrange the numbers in order from least to greatest.

2. Give your numbers to a partner to arrange in order from least to greatest.

3. Check your partner's answers against your own. If there are any differences, check both your own answers and your partner's again to see where the errors were.

4. Write a sentence about each pair of numbers. The first one is done for you:
   a) 56 880 and 97 456 : 97 456 is greater than 56 880.
   b) 80 782 and 80 781
   c) 1 008 752 and 1 000 000
   d) 12 709 and 99 009

### Math Problems to Solve

5. Write a word problem that uses large numbers. Solve your problem, and then give it to a classmate to answer.

6. After your partner has solved your problem, have him or her complete an evaluation of your problem, pointing out things that worked and things that didn't work or that weren't clear.

7. Improve your problem based on your partner's feedback.

**Technology**

Use a desktop publishing program to create flashcards. Each flashcard should contain a large single digit from 0 to 9. Create enough flashcards to make 2 or 3 of each digit. Print out your sheets of flashcards and cut them out. Use your flashcards to create different seven-, eight-, and nine-digit numbers. Practise saying the numbers aloud, order them from least to greatest, and ask a classmate questions so that he or she can guess which number you are holding (without seeing the number).

**Lesson 3**

# Finding Missing Values

SPACE LOG

PLAN:
You will solve for a missing value while studying comets and meteors.

DESCRIPTION:
Comets are bits of ice, dust, and rock. When a comet gets close to the sun, the ice starts to melt, forming a long tail. A comet's tail can be millions of kilometres long!

Meteors are bits of comet dust that have entered Earth's atmosphere. We call them "shooting stars."

## Get Started

Like the planets, comets orbit the sun.

**1.** Copy this chart into your notebook, and complete the last column of the chart.

| Comet | Year Last Seen | Approximate Orbital Period (years) | Year Next Seen |
|---|---|---|---|
| Halley | 1986 | 76 | |
| Encke | 2000 | 3 | |
| Faye | 1999 | 7 | |
| Pons-Brooks | 1954 | 70 | |
| Crommelin | 1984 | 27 | |
| Swift-Tuttle | 1992 | 135 | |
| Tuttle | 1994 | 14 | |
| Kojima | 2000 | 7 | |

**2.** Which comets will return in your lifetime?

**3.** Which comets passed by Earth before you were born?

**4.** How did you find the answers?

## Build Your Understanding

### Solve Equations

Halley's comet will be visible from Earth in 2062. It was last seen in 1986. What is its orbital period?

### Vocabulary

**equation:** A mathematical statement where each side equals the other. The sides are separated by an equals sign.

**1.** We can write this problem as 2062 − 1986 = ■. (This is called an equation.) The missing value in this equation is the orbit period. It is represented with a ■.

### Journal

List the steps you would follow to solve an equation like 1986 + ■ = 2062.

We can also write the equation as 1986 + ■ = 2062.

**2.** Here is one way to solve the second equation. Follow these steps:

**a)** 1986 + ■ = 2062    **b)** ■ = 2062 − 1986

**c)** ■ = 76

**3.** Explain why subtraction helped to solve the addition equation.

**4.** Write equations for all of the comets in the chart below:

| Comet | Year Last Seen | Approximate Orbital Period (years) | Year Next Seen |
|---|---|---|---|
| Howell | 1998 | | 2004 |
| Wolf-Harrington | | 7 | 2004 |
| Tempel-Tuttle | 1998 | | 2031 |
| Slaughter-Burnham | 1993 | | 2005 |
| 2000 WT168 | | 7 | 2008 |
| Kowal 1 | 1992 | | 2007 |
| Shoemaker-Levy 3 | | 7 | 2005 |
| Brorsen-Metcalf | 1989 | | 2059 |

**5.** Solve the equations to determine the missing value. Show your steps.

1. Rewrite the problems from Get Started as missing-value equations using addition. The first two have been done for you:

   **a)** 1986 + ■ = 2062

   **b)** 2000 + ■ = 2003

2. Explain how to solve an equation like the ones in this lesson's Build Your Understanding using numbers, pictures, and words.

3. Compare your answer for question 2 with a classmate's.

**Practice**

Answer these questions mentally:

**1.** 2365 + ■ = 3497      **2.** 362 + ■ = 4562      **3.** 3634 − ■ = 2134

**4.** 56 + ■ = 369      **5.** 45 − ■ = 37      **6.** 3621 − ■ = 234

**7.** 363 + ■ = 434      **8.** 3631 − ■ = 298

Answer the following questions:

**9.** 18 x 2 = ■      **10.** 12 x 12 = ■      **11.** 18 − 9 = ■

**12.** 40 − 9 = ■      **13.** 10 + 157 = ■      **14.** 102 x 5 = ■

**15.** 4 x 8 = ■      **16.** 425 − 15 = ■      **17.** 336 − 118 = ■

**18.** 420 + 20 = ■

Write new equations for the questions above so that the missing value is for a different number. The first one has been done for you.

**19.** 18 x ■ = 36

**Extension**

20. Encke is a comet that orbits the sun once every 3 years. Kojima is a comet that has a 7-year orbital period. They both were last seen from Earth in the year 2000. When will both comets appear again in the same year? Show your work.

## Show What You Know

1. Make up six large numbers using the digits 0 to 9. Arrange the numbers from least to greatest. Write each of your numbers in words and in expanded form. Give your answers to a classmate to check.

2. How much farther is Mars from the sun than Mercury is from the sun, Venus is from the sun, and Earth is from the sun? Estimate first, and then solve.

3. If Earth takes 365 days to orbit the sun, Mercury takes 88 days to orbit the sun, and Venus takes 255 days to orbit the sun, how old could you be on each planet?

4. How often will the comets Tuttle, Crommelin, and Kojima be seen from now until 2050? Write an equation showing the pattern.

## Lesson 4

# Solving Equation Problems

SPACE LOG

PLAN:
You will solve missing-value equations involving space probes and space exploration.

DESCRIPTION:
The first space probe sent from Earth reached Venus in 1962. Since then, dozens of probes have been sent to explore deep space, far beyond where humans have travelled. On September 5, 1977, Voyager 1 was launched. It arrived at Jupiter in March of 1979.

**Voyager 1 spacecraft**

## Get Started

1. Solve these equations mentally:

   **a)** $14 + \blacksquare = 28$  **b)** $36 + \blacksquare = 60$

   **c)** $5 \times \blacksquare = 30$  **d)** $9 \times \blacksquare = 27$

   **e)** $4 \times \blacksquare = 32$  **f)** $24 \div \blacksquare = 8$

   **g)** $36 \div \blacksquare = 6$  **h)** $56 \div \blacksquare = 8$

   **i)** $72 - \blacksquare = 60$  **j)** $99 - \blacksquare = 33$

### Booklink

**Galileo Spacecraft: Mission to Jupiter** by Michael D. Cole (Enslow Publishers, Springfield, NJ, 1999). This book is about the *Galileo* spacecraft from its launch to its orbit around Jupiter.

2. Solve these equations in your notebook. Show your steps.

   **a)** $3001 + \blacksquare = 4785$  **b)** $984 + \blacksquare = 1001$

   **c)** $805 - \blacksquare = 419$  **d)** $5243 - \blacksquare = 4320$

   **e)** $48 \times \blacksquare = 1152$  **f)** $24 \times \blacksquare = 576$

   **g)** $14 \times \blacksquare = 56$  **h)** $48 \times \blacksquare = 240$

   **i)** $120 \div \blacksquare = 24$  **j)** $96 \div \blacksquare = 48$

## Write Equations

The space probes sent from Earth travel at great speeds, yet they still take many, many years to reach their destinations. It took the spacecraft *Galileo* over six years to reach Jupiter after it was launched from the space shuttle.

For each problem below, what value is missing? Write an equation using a ■ to represent the missing value. Then solve each problem using the equation you made. You may use a calculator to help you.

1. A probe is being launched from the next space shuttle mission to explore the rings of Saturn. If it travels at 1400 km per hour, how many km will it travel in 24 h?

2. If a probe takes 6 years to travel to Jupiter, a distance of 778 000 000 km, how many kilometres does it travel in 1 month?

3. A NASA mission to Mars is expected to take 6 months just to travel to the planet. When Earth and Mars are closest, the distance between them is 55 000 000 km. How many km will the spacecraft travel in 1 month on its mission to Mars? How many km will the spacecraft travel in 4 months?

## What Did You Learn?

1. What helped you to write the equations in this lesson's Build Your Understanding?

2. Compare one of your problem-solving equations with a classmate's. Did you follow the same steps and develop the same equations? Why or why not?

3. Draw a picture to explain each problem in this lesson's Build Your Understanding. Do you find it easier to explain your answer to each problem using numbers, pictures, or words? Explain your choice.

### Technology

Add an entry explaining the rule for solving missing-value equations for addition, subtraction, multiplication, and division to your personal reference guide that you made in Chapter 1. Don't forget to save the updated version.

Solve the following equations mentally:

**1.** 4 x ▦ = 36　　**2.** 3 x ▦ = 45　　**3.** ▦ x 8 = 56

**4.** ▦ − 7 = 32　　**5.** 63 ÷ ▦ = 9　　**6.** ▦ + 104 = 114

Solve these equations. Share your answers with a classmate.
Why might your answers be different?

**7.** ▲ x ▦ = 48　　**8.** ▲ x ▦ = 24　　**9.** ▲ x ▦ = 12

Solve the following equations in your notebook. Show your steps.

**10.** ▦ + 4678 = 8190　　**11.** 3599 − ▦ = 2854

**12.** 76 x ▦ = 3268　　**13.** ▦ x 456 = 15 504

**14.** 248 ÷ ▦ = 4　　**15.** ▦ ÷ 327 = 3

**16.** 350 ÷ ▦ = 5　　**17.** ▦ x 2 = 46

**18.** ▦ x 10 = 2800　　**19.** 29 x ▦ = 174

**20.** 5723 + ▦ = 5810　　**21.** 850 − ▦ = 424

**22.** ▦ x 25 = 17 225　　**23.** ▦ ÷ 5 = 1305

## Extension

**24.** *Voyager 1* was launched in September 1977. After 18 months of travelling through space, it reached Jupiter. Another 20 months later, it reached Saturn. In what month and year did *Voyager 1* reach Saturn?

**Tip**

Rewrite important information from this question.
Look for key words to help you.

**Lesson 5**

# Solving for Variables

SPACE LOG

PLAN:
You will compare times on Earth and
Jupiter using equations that have letters
in place of the missing value.

DESCRIPTION:
Jupiter is the largest planet in our solar
system. It also rotates on its axis faster
than any other planet. A day on Jupiter
is only 9.8 Earth hours long!

## Get Started

Li has many questions about Jupiter.
She decides to organize her questions
as math problems, which are recorded
below. For each question, what value
is missing? Write an equation using ■
to represent the missing value. Solve
the equation. You may use a calculator
to help you. Then, write a new
equation so that the missing value is
for a different number. What does ■
represent in your new equation? The
first one has been done for you.

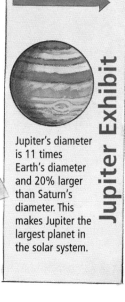

Jupiter's diameter
is 11 times
Earth's diameter
and 20% larger
than Saturn's
diameter. This
makes Jupiter the
largest planet in
the solar system.

**Jupiter Exhibit**

1. If 1 day on Jupiter is 9.8 hours long,
   how many hours are there in 4 days?
   The missing value is the number of
   hours in 4 days.

   9.8 x ■ = ■

   9.8 x 4 = 39.2

   9.8 x ■ = 39.2

   The new missing value is the
   number of days.

**2.** If 1 day on Jupiter is 9.8 hours long, how many Jupiter days are in 117.6 hours?

**3.** If 1 day on Jupiter is 9.8 hours long, how many Jupiter days are in 3577 hours?

**4.** If 1 day on Jupiter is 9.8 hours long, how many hours are in 48 Jupiter days?

## Build Your Understanding

### Write Equations

It takes Jupiter almost 12 years to orbit the sun. That makes 1 Jupiter year almost 12 times longer than 1 Earth year. Written as an equation, it looks like this:

1 x 12 = 12

If you write the equation with a missing value representing Jupiter's orbit, it looks like this:

1 x ▪ = 12

Sometimes we put a letter in place of the box. It looks like this:

1 x $n$ = 12

In this equation, $n$ represents Jupiter's orbit. A letter used to represent something in an equation is called a variable.

To solve this problem, think

$n$ = 12 ÷ 1

$n$ = 12

Why does division help to solve a multiplication equation?

Solve these problems.

**1.** 10 x $n$ = 10    **2.** 2 x $n$ = 10    **3.** 33 x $n$ = 99    **4.** 4 x $n$ = 160

**5.** 4 x $n$ = 8    **6.** 2 x $n$ = 12    **7.** 3 x $n$ = 45    **8.** 8 x $n$ = 240

**9.** 9 x $n$ = 27    **10.** 4 x $n$ = 320

**Journal**

List and explain the steps you follow when you write an equation.

**Vocabulary**

**variable:** A letter or symbol that is used in an equation to represent an unknown or missing value

1. What does the letter $n$ stand for in the equations in Build Your Understanding?

2. Think of a rule for solving a missing-value addition equation such as $45 + n = 101$. Now think of a rule for solving a missing-value subtraction question such as $n - 2 = 10$, a missing-value multiplication question such as $5 \times n = 35$, and a missing-value division question such as $n \div 3 = 12$.

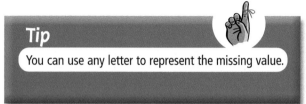

**Tip**

You can use any letter to represent the missing value.

## Practice

1. Write an equation for this problem: If 1 Jupiter day is 9.8 hours, how many Jupiter days are there in 1 Earth year?

Solve these equations mentally:

| | | |
|---|---|---|
| **2.** $2 \times n = 4$ | **3.** $3 \times n = 12$ | **4.** $5 \times n = 20$ |
| **5.** $9 \times n = 18$ | **6.** $4 \times n = 16$ | **7.** $7 \times n = 14$ |
| **8.** $10 \times n = 100$ | **9.** $2 \times n = 20$ | **10.** $3 \times n = 60$ |

Solve these equations in your notebook. Show your steps.

| | | |
|---|---|---|
| **11.** $10 \times n = 1210$ | **12.** $2 \times n = 48$ | **13.** $30 \times n = 900$ |
| **14.** $48 \times n = 96$ | **15.** $75 \times n = 225$ | **16.** $n \div 4 = 16$ |
| **17.** $n \div 10 = 4$ | **18.** $n \div 7 = 8$ | **19.** $n \div 10 = 10$ |
| **20.** $n \div 125 = 2$ | **21.** $500 \div n = 250$ | **22.** $36 \times n = 180$ |

Solve these equations in your notebook:

| | | |
|---|---|---|
| **23.** $n - 16 = 6$ | **24.** $n + 50 = 67$ | **25.** $n - 23 = 52$ |
| **26.** $n + 82 = 89$ | **27.** $n - 7 = 114$ | **28.** $n + 307 = 480$ |

## A Math Problem to Solve

**29.** Eric has 3 days to finish his math homework. If Eric lived on Jupiter, approximately how many Earth days would he have to complete his homework? Would he have more time or less? Explain your answer and show your work using numbers, pictures, and words.

**Lesson 6**

# More Missing-Value Equations

SPACE LOG

PLAN:
You will study problems about Saturn and solve missing-value equations.

DESCRIPTION:
Saturn is the second-largest planet in our solar system. It rotates almost as fast as Jupiter, once every 10.665 hours. This makes a day on Saturn 10.665 Earth hours long. Saturn takes 29.5 Earth years to orbit the sun.

## Get Started

Saturn takes 29.5 years to orbit the sun, but once every year and 14 days, Earth, Saturn, and the sun lie in a direct line. This is called the opposition. When they are in opposition, Earth and Saturn are the closest they will be to each other for another year and 14 days.

If Saturn is closest to Earth on January 2 one year, what day will it be closest the following year? What about the year after that?

## Solve Complex Equations

If Saturn is closest to Earth on Day 1 of this year, on what day will it be closest 4 years from now?

You can solve this problem in two ways:

**A.** Tony figured out each year in turn.

This year = Day 1

1 year later = Day 1 + 14 days = Day 15

2 years later = Day 1 + 14 days + 14 days = Day 29

3 years later = Day 1 + 14 days + 14 days + 14 days = Day 43

4 years later = Day 1 + 14 days + 14 days + 14 days + 14 days = Day 57

**B.** Mahalia wrote an equation.

Day 1 + $n$ x 14 = Day of next opposition

In this equation, $n$ represents the number of years.

To solve this equation, Mahalia let $n = 4$.

1 + 4 x 14 =

1 + 56 =

= 57

Use the equation to find out on what day of the year Saturn and Earth will be closest to each other 10 years from now.

## What Did You Learn?

1. Why is using an equation easier than figuring out this problem year by year?

2. Think of another problem in which an equation might be helpful.

3. Explain how Mahalia figured out the equation "Day 1 + $n$ x 14 = Day of next opposition (where $n$ represents the number of years)," from Tony's method of calculation.

## Practice

Solve these equations if $n = 5$:

1. $3 + n = $
2. $1 + n \times 2 = $
3. $3 + n \times 4 = $
4. $15 \div n - 2 = $
5. $8 \times n - 4 = $
6. $6 \times n + 3 = $

> **Tip**
>
> Remember to use the correct order of operations to solve equations.

7. Solve the same equations as in questions 1 to 6, but this time $n = 3$.

### Math Problems to Solve

8. Create a word problem that involves a missing value. Create the equation your problem requires and insert a variable for the missing value. Using your equation, solve the problem.

9. Now give your word problem to a classmate to solve. Check his or her equation and answer against your own.

### Show What You Know

**Review: Lessons 4 to 6, Problem Solving**

1. Imagine that a space probe travels 1400 km/h and it is going to Mars from Earth. From the data in Lesson 4, calculate how far the space probe would travel in 1 year and write an equation for this problem.

2. Jupiter takes 12 years to orbit the sun, which makes 1 Jupiter year almost 12 times longer than 1 Earth year. How many Earth years would it take you to become 24 years old on Jupiter? Write an equation for this problem.

3. If 1 day on Saturn is 10.665 h, how many hours would 365 days be on Saturn? Compare this amount to the amount of hours 365 days would be on Earth. Write equations for these problems.

## Lesson 7

# Analyzing Data

*SPACE LOG*

*PLAN:*
**You will conduct a survey about Uranus and evaluate the results.**

*DESCRIPTION:*
Uranus is so far from Earth that we do not know a lot about this planet. It is the seventh planet from the sun in our solar system. Uranus has at least 21 moons. It also has a greenish-coloured atmosphere that is made up of 83% hydrogen, 15% helium, and 2% methane.

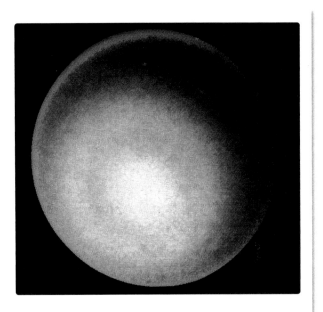

## Get Started

Conduct a survey to find out how much people know about the planet Uranus. You may work with a partner.

**1.** Design your survey. What type of information do you want to find out? What questions will you ask? How will you record your results? Whom will you ask?

**2.** Conduct your survey. Record your results.

**3.** Analyze your results. What do they tell you?

**4.** Based on your results, what sort of information do you think should be included in a "Planet Uranus" presentation?

### Vocabulary

**survey:** A method of gathering information by asking questions and recording people's answers

## Review Data

Here is a chart of information about some of Uranus' moons.

| Uranus' Moons | | | |
|---|---|---|---|
| Name | Diameter (km) | Distance From Uranus (km) | Year Discovered |
| 1986 U10 | 40 | 75 000 | 1999 |
| Ariel | 1160 | 191 240 | 1851 |
| Belinda | 66 | 75 260 | 1986 |
| Bianca | 42 | 59 000 | 1986 |
| Caliban | 80 | 7 200 000 | 1997 |
| Cordelia | 26 | 49 750 | 1986 |
| Cressida | 62 | 61 770 | 1986 |
| Desdemona | 54 | 62 660 | 1986 |
| Juliet | 84 | 64 360 | 1986 |
| Miranda | 472 | 129 780 | 1948 |
| Oberon | 1526 | 582 600 | 1787 |
| Ophelia | 30.4 | 53 440 | 1986 |
| Portia | 108 | 66 085 | 1986 |
| Prospero | 30 | 16 568 800 | 1999 |
| Puck | 154 | 86 010 | 1985 |
| Rosalind | 54 | 69 941 | 1986 |
| Setebos | 30 | 17 681 000 | 1999 |
| Stephano | 20 | 7 948 000 | 1999 |
| Sycorax | 160 | 12 200 000 | 1997 |
| Titania | 1580 | 435 840 | 1787 |
| Umbriel | 1190 | 265 970 | 1851 |

**1.** Review the data on this chart.

**2.** Put the moons in order from smallest to largest. What do you notice about the size of the moons?

**3.** Put the moons in order from closest to Uranus to farthest from Uranus. What do you notice?

**4.** What is the mean diameter of Uranus' moons? Use a calculator to help you.

**5.** What is the mean distance of the moons from Uranus? Use a calculator to help you.

**6.** What is the range of the years in which Uranus' moons were discovered?

### Vocabulary

**mean:** The average of a set of amounts. To calculate the mean, add up all the amounts and divide the sum by the total number of addends given. The mean of 2, 3, 5, 6, and 9 is 5.

**range:** The difference between the greatest and least values in a group. For example, the youngest student in the school is 5, and the oldest student in the school is 13. 13 − 5 = 8. Therefore, the range is 8.

### Technology

After you complete the Build Your Understanding section, complete it again using a spreadsheet computer program. Use the "sort" tool to order the moons' diameters. You can also use the "average" tool to find the mean. Which method did you prefer to complete the activity — by hand or by computer? Why?

## What Did You Learn?

**1.** How does reviewing data about Uranus answer any questions you might have had about the planet?

**2.** Does the data cause you to have more questions? Make a list of those questions.

**3.** Why is reading information in a chart useful?

## Practice

1. Make a bar graph of the diameter column from the Uranus' Moons chart in this lesson's Build Your Understanding. Don't forget to label both axes, and to give your graph an appropriate title.

2. Now make a line graph of the number of moons discovered each year. List the years in order on the horizontal axis, from earliest to most recent discovery.

3. List anything interesting that you notice about both graphs that you made.

## Extension

4. Is there a relationship between the size of the moon and the distance it is from the planet Uranus? Use a computer program to create a scatterplot of the data to help you decide.

5. Design a survey about any of the other planets.

6. Conduct your survey and compare the results to those of your Uranus survey.

### Vocabulary

**bar graph:** A graph that uses parallel bars to show the relationship between quantities

**line graph:** A graph that uses a line to show how data changes over a period of time

**scatterplot:** A graph that uses plotted dots or points to show the relationship between two variables

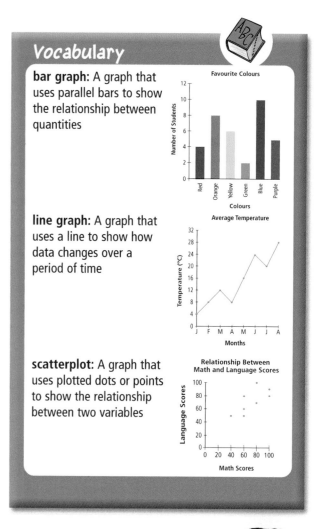

## Lesson 8

# Comparing Data

*SPACE LOG*

*PLAN:*
**You will compare sets of data about different planets' moons.**

*DESCRIPTION:*
Neptune has eight moons that we know of. Seven are very small, but the eighth, Triton, is much larger than the others. What other planets have moons besides Earth and Neptune? How many moons do each of them have?

**Triton, Neptune's largest moon**

## Get Started

Look at the data on this chart.

| Planet | Number of Known Moons | Planet | Number of Known Moons |
|--------|----------------------|--------|----------------------|
| Mercury | 0 | Saturn | 30 |
| Venus | 0 | Uranus | 21 |
| Earth | 1 | Neptune | 11 |
| Mars | 2 | Pluto | 1 |
| Jupiter | 52 | | |

1. What is the mean number of moons per planet? How is this a confusing number?

2. Graph this data. What type of graph would be best to display this data?

**Technology**

Use a spreadsheet computer program to create your chart and graph from Get Started.

**Vocabulary**

**mean:** The average of a set of amounts. To calculate the mean, add up all the amounts and divide the sum by the total number of addends given.

## Build Your Understanding

## Compare Data

1. Look at the data in the chart below.

| Neptune's Moons | | | |
|---|---|---|---|
| Name | Diameter (km) | Distance From Neptune (km) | Year Discovered |
| Despina | 148 | 53 000 | 1989 |
| Galatea | 158 | 62 000 | 1989 |
| Larissa | 192 | 74 000 | 1989 |
| Naiad | 58 | 48 000 | 1989 |
| Nereid | 340 | 5 509 000 | 1949 |
| Proteus | 418 | 118 000 | 1989 |
| Thalassa | 80 | 50 000 | 1989 |
| Triton | 2700 | 355 000 | 1846 |

2. Compare the data in the chart above to the data in Lesson 7 about the moons of Uranus on page 128. List the similarities and differences in your notebook.

3. Make a bar graph of the diameter column from the chart above. Make a line graph of the number of moon discoveries each year. List the years in order on the horizontal axis from earliest to most recent discovery.

**4.** Compare your graphs with the two graphs you made in Lesson 7. What do the graphs tell you?

**5.** Create five questions for a classmate to answer about both the graph of Uranus' Moons (in Lesson 7) and the graph of Neptune's Moons.

## What Did You Learn?

**1.** What information do you find when you compare the data in the two charts and the graphs?

**2.** How is this information helpful?

## Practice

Find the mean of the following sets of numbers:

**1.** 34, 56, 89, 23, 2, 98, 17

**2.** 3, 107, 56, 87, 12, 31, 66

**3.** 57, 136, 582, 3, 90, 123

**4.** If the mean of 5 numbers is 81, what might the numbers be?

**5.** If the mean of 7 numbers is 5, what might the numbers be?

### Extension

Choose another topic about space and find data on it.

**6.** Research your space topic, perhaps on the Internet (with your teacher's permission), or from books in the library.

**7.** Graph your data. Choose an appropriate type of graph and be prepared to explain your choice.

## Lesson 9

# Interpreting Data

SPACE LOG

PLAN:
You will interpret data about the orbits of the planets.

DESCRIPTION:
Pluto is the farthest planet from the sun and has the longest orbital period of all the planets. It has one moon, which is almost half its size. Because of this, sometimes Pluto and its moon, Charon, are called a "double planet."

## Get Started

1. Use the Internet or the library to research data about Pluto.

2. What does the data tell you about Pluto?

3. How helpful is the data?

4. Would the data be more useful if you knew similar data about the other planets? Why or why not?

## Build Your Understanding

### Interpret Data

1. Look at the data in this chart. What information does it give you?

2. Write statements comparing each planet to another. For example, you could write: Earth travels through space about six times faster than Pluto.

| Planet | Orbital Speed (km/s) |
|--------|----------------------|
| Mercury | 47.9 |
| Venus | 35.0 |
| Earth | 29.8 |
| Mars | 24.1 |
| Jupiter | 13.1 |
| Saturn | 9.6 |
| Uranus | 6.8 |
| Neptune | 5.4 |
| Pluto | 4.7 |

**3.** Look at the information in the chart to the right and compare it to the previous chart.

**4.** What new statements can you make about each planet based on what these two charts tell you? How does the information in one chart help explain the information in the other chart?

| Planet | Orbital Period |
|---|---|
| Mercury | 88 days |
| Venus | 225 days |
| Earth | 365 days |
| Mars | 687 days |
| Jupiter | 11.9 years |
| Saturn | 29.5 years |
| Uranus | 84 years |
| Neptune | 165 years |
| Pluto | 248 years |

## What Did You Learn?

**1.** Look at your statements and comparisons. Explain how the data led you to each statement and comparison.

**2.** How can you prove that your statements are right?

**3.** Could other types of data have helped you make more accurate statements? What types?

### Booklink

**Incredible Comparisons** by Russell Ash (Houghton Mifflin: New York, NY, 1996). This book is a collection of facts and comparisons on many different topics, including big buildings, animal speeds, growth and age, and population.

### Technology

Use a spreadsheet computer program to create a line graph that displays the data from both charts in this lesson's Build Your Understanding. (Remember that in order to create the line graph, you must first enter the data into a spreadsheet.)

1. Survey your classmates on a topic that interests you.

2. Organize the results in a chart like the one in this lesson's Build Your Understanding.

3. Write three statements comparing the data in the chart.

4. Write three questions about the chart for a classmate to answer.

5. Compare your statements to a classmate's. How are they similar and different?

6. Find the mean of the orbital speeds for all the planets.

7. Now find the mean of the orbital periods for all the planets. Round your answer to the nearest tenth. You may use a calculator to help you.

## Extension

8. Graph the data from your survey in Practice question 1. Choose an appropriate type of graph and explain why you chose the graph you did. Share your graph with the class.

9. List the steps you followed to graph your survey results.

10. How else is data important in exploring space?

## Lesson 10

# Mean, Median, and Mode

*SPACE LOG*

*PLAN:*
You will calculate the mean, median, and mode for the diameters of the nine planets.

*DESCRIPTION:*
The students have seen all they can in one day at the space exhibit in the Science Centre. They head back to their school to think about all they have learned about math and exploring space.

## Get Started

Compare the diameter of Earth to the diameter of the other planets. There are as many planets that are smaller than Earth as there are planets that are larger. What are the four planets that are smaller than Earth? Which four are larger? Look at the chart to the right to help you.

| Planet | Diameter (in km) |
| --- | --- |
| Mercury | 4 880 |
| Venus | 12 100 |
| Earth | 12 756 |
| Mars | 6 794 |
| Jupiter | 143 000 |
| Saturn | 120 500 |
| Uranus | 51 100 |
| Neptune | 49 500 |
| Pluto | 2 300 |

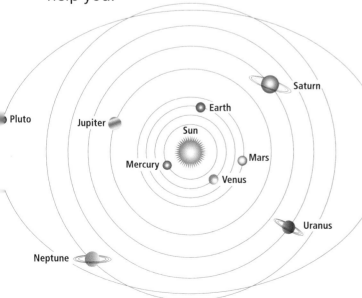

When numbers are listed from least to greatest, the number in the middle is called the median. Therefore, Earth is the median planet with respect to its size.

## Build Your Understanding

### Compare Mean, Median, and Mode

Look at the chart in Get Started to answer the following questions:

1. Calculate the mean diameter of the nine planets. Use a calculator to help you.

2. Is there a mode in this data? How do you know?

3. What is the median?

4. How do these three numbers compare?

5. How would you find the median in a set of numbers that has two middle values? Share and compare your answer with the class.

## What Did You Learn?

1. Which answer would you use (from question 1, 2, or 3 in this lesson's Build Your Understanding) if you were asked to find the average diameter of the planets in our solar system? Why?

2. Think of some instances where calculating the mean would be useful in analyzing data. Now think of some instances where calculating the mode, and then the median, would be useful in analyzing data. Record your ideas in your notebook.

## Practice

1. Find the mean, median, and mode for each of these groups of numbers:

   a) 466, 485, 555, 767, 466, 201

   b) 18, 24, 18, 24, 66, 99, 24

   c) 100, 200, 100, 600, 700, 900, 400

Look at this set of data on math scores:

| Student | Score out of 20 |
|---------|-----------------|
| Ahanu | 16 |
| Li | 15 |
| Eric | 19 |
| Ruth | 18 |
| Phillip | 15 |
| Tony | 20 |
| Claire | 19 |
| Mahalia | 19 |
| Joey | 16 |

**Technology**

Add definitions for mean, median, and mode to your personal reference guide that you started in Chapter 1. Don't forget to show some examples. Remember to save the updated version.

**2.** What is the mean score?

**3.** What is the mode of these scores?

**4.** What is the median score?

**5.** Which of the three terms best describes how the students scored as a whole group?

**6.** Write a statement about the planets' diameters using the mean data that you calculated in question 1 of Build Your Understanding.

**7.** Write a statement about the planets' diameters using the median data that you calculated in question 3 of Build Your Understanding.

## Lesson 11

# Describing Data

SPACE LOG

PLAN:
You will describe data about space beyond Pluto.

DESCRIPTION:
The Grade 6 math class knows there is much more left to explore beyond Pluto. There are stars and nebulae, pulsars and black holes, quasars and galaxies. To explore it all, math is a tool they will need to use.

The Pleiades star cluster is one of the brightest star clusters visible in the northern hemisphere.

## Get Started

There are hundreds of billions of stars in the Milky Way, and the Milky Way is only one of several hundred million galaxies. But of all those stars, only 8000 are visible from Earth. From North America, we can see half of those (about 4000). And half of those we can't see because they aren't visible during daylight hours. That means there are only 2000 stars to look at every night.

From 2000 to hundreds of billions is a vast range of numbers. What would be the best way to present data that has such a wide range? Work with a small group to brainstorm ways and present your findings to the rest of the class.

### Describe Data

The table on the right shows the 10 stars closest to Earth.

1. Put the distances from Earth in order from closest to farthest.

2. Calculate the mean distance of these stars from Earth.

3. What is the median distance?

4. What is the mode distance?

| Star | Distance From Earth (light years) |
| --- | --- |
| Alpha Centauri A | 4.3 |
| Alpha Centauri B | 4.3 |
| Barnard's Star | 6.0 |
| Lalande 21185 | 8.2 |
| Proxima Centauri | 4.3 |
| Sirius A | 8.6 |
| Sirius B | 8.6 |
| UV Ceti A | 8.4 |
| UV Ceti B | 8.4 |
| Wolf 359 | 7.8 |

### Vocabulary

**light year:** The distance light can travel in one Earth year. Light travels at 297 600 km per second, thus, 1 light year is 9 404 800 000 000 km!

## What Did You Learn?

1. The mean, median, and mode you calculated in Build Your Understanding are for the 10 stars closest to Earth. If the chart included stars that aren't necessarily close to Earth, how would your results change?

2. Which did you find easiest to calculate: the mean, the median, or the mode? Explain your answer.

### Tip

Here's another NASA Web site. It has all sorts of information about the universe beyond our solar system:
http://starchild.gsfc.nasa.gov/docs/StarChild/StarChild.html

### Booklink

**Touching the Earth** by Roberta Lynn Bondar (Key Porter Books: Toronto, ON, 1994). Roberta Bondar describes her mission in space to the advanced reader. She also talks about how her understanding of Earth has changed since her travels in space.

What is the median for the following sets of numbers?

**1.** 189, 34, 576, 1234, 256, 978

**2.** 32, 9, 2, 12, 6, 24, 19, 21, 16, 18

**3.** 1634, 136, 367, 956, 790, 512, 1256, 1411

**4.** Check your answers with a classmate's.

Calculate the mean, median, and mode of these sets of numbers:

**5.** 45, 62, 79, 85, 105, 85, 62, 85

**6.** 21, 152, 2031, 31, 2, 0, 5, 21

**7.** 2, 9, 8, 8, 6, 8, 45, 62, 17, 8

Solve the following equations based on the data provided in Get Started:

**8.** $8000 - n - 2000 = 2000$

**9.** $2000 + 2000 + n = 8000$

**10.** $n - 4000 + 4000 = 8000$

## Extension

**11.** Create a word problem for a classmate to solve using the data provided in this lesson.

## Show What You Know

**Review: Lessons 7 to 11, Surveys, Graphs, Mean, Median, and Mode**

**1.** Design a survey to find out what foods the students in your class prefer. Have students select from 5 food items. Display your information in a graph.

**2.** Find the mean price of the following food items:

**a)** pop: $1.25

**b)** chips: $1.15

**c)** granola bar: $1.39

**d)** bag of cookies: $3.99

**e)** ice cream: $1.99

**f)** box of cereal: $4.99

If you had $8.00 to spend, show all the different combinations that you could purchase without going over.

**3.** Calculate the mean, median, and mode for the following batting averages:

250, 265, 300, 295, 315, 272, 185, 215, 295

Graph this data.

# Chapter Review

**1.** Write a sentence comparing each pair of numbers.

**a)** 169 001   87 541 362      **b)** 16 127 323   954 446 126

**c)** 13 806 257   450 321      **d)** 24   240 000

**2.** Write these numbers on a place-value chart:

**a)** 26 131 701      **b)** 323 456 113      **c)** 873 205 562

**d)** 1 880 349      **e)** 66 180 310      **f)** 24 709 003

**3.** Rewrite these equations so that they have a missing value:

**a)** 4 x 8 = ▣      **b)** 10 x 10 = ▣      **c)** 21 − 7 = ▣

**d)** 45 − 5 = ▣      **e)** 7 + 20 = ▣      **f)** 5 x 4 = ▣

**g)** 2 x 11 = ▣      **h)** 23 − 11 = ▣      **i)** 16 − 12 = ▣

**j)** 10 + 8 = ▣

**4.** Solve these equations. Show your steps.

**a)** 2163 + ▣ = 4670      **b)** ▣ − 227 = 645

**c)** 84 ÷ ▣ = 7      **d)** 11 x ▣ = 1100

**e)** 470 + ▣ = 567      **f)** 147 ÷ ▣ = 21

**g)** 13 x ▣ = 156      **h)** 6704 − ▣ = 2075

**i)** 8 x ▣ = 112      **j)** 11 x ▣ = 220

**k)** ▣ ÷ 16 = 22      **l)** ▣ ÷ 8 = 56

**5.** Solve these equations:

**a)** 8 x $n$ = 512      **b)** 4 x $n$ = 12

**c)** 18 x $n$ = 36      **d)** 25 x $n$ = 175

**e)** 13 x $n$ = 52      **f)** 21 ÷ $n$ = 7

**g)** ▣ ÷ 2 = 102      **h)** ▣ ÷ 11 = 4

**i)** ▣ ÷ 14 = 2      **j)** ▣ ÷ 5 = 19

**6.** Find the mean, median, and mode for each of these groups of numbers:

    **a)** 133, 405, 78, 77, 64, 133

    **b)** 54, 99, 108, 7761, 445, 446, 445

    **c)** 1000, 5000, 6000, 1000, 1000, 6000, 9000, 1000, 8000

**7. a)** Make a bar graph of the information in this tally chart:

| Favourite Planets | |
|---|---|
| Planets | Number of Students |
| Mars | 卌 \|\|\|\| |
| Earth | 卌 卌 卌 |
| Mercury | \|\|\|\| |
| Venus | \|\|\|\| |
| Jupiter | 卌 \|\|\|\| |
| Saturn | \|\|\|\| |
| Uranus | 卌 \|\|\|\| |
| Neptune | \|\|\|\| |
| Pluto | \|\|\|\| |

    **b)** What is the mode number?

    **c)** What is the median number?

    **d)** What is the mean number?

    **e)** Write three statements comparing the information in your graph.

# Chapter Wrap-Up

You have reached the end of Chapter 3. In this chapter, you have learned about math beyond Mars.

For this project, you are going to plan and present research about a planet or star of your choice.

## Research

You need to decide what you will research and where you will find out what you need to know. Will you use your school library, the public library, the Internet, the local planetarium, or a combination of these?

## Present Your Research

You need to decide how to present your research. You can present a report, design a poster, make a 3-D model, or give a multimedia presentation. In your presentation you must show how math is needed to research and explore the planet or star that you choose to study. Be sure to include all of the mathematical concepts learned in this chapter.

Good luck!

### Technology

Present your project using a computer program of your choice.

# Problems to Solve

Here are some fun problems for you to solve. For each problem, a helpful problem-solving strategy is included for you to use. Later in the year, once you have learned more strategies, you will have the chance to choose the strategies you want to use.

## Problem 6

# Music Survey

STRATEGY: SOLVE A SIMPLE OR SIMILAR PROBLEM

If you find that you are working on a very difficult problem, you can make it easier to solve by changing it to a simpler problem. For instance, you can change large numbers to smaller numbers or reduce the number of items in a problem. You might also want to look for a pattern or decide on an easier way to find the solution.

OBJECTIVE:

Demonstrate an understanding of percentages

## Problem

**You Will Need**
• 1-cm grid paper

**1.** You conducted a survey in your class to see who liked rap music. Of the 35 students, 60% indicated that they liked rap. How many students did not like rap music?

**Tip**

**Problem-Solving Steps**
1. Understand the problem
2. Pick a strategy ("Solve a Simple or Similar Problem")
3. Solve the problem
4. Share and reflect

**2.** In another survey that you conducted, the students who did not like rap music were asked which type of music they liked the most: jazz music, pop music, or alternative music. The number of students who chose jazz music was equal to the number who chose pop music. 4 students chose alternative music. How many students chose each type of music?

**3.** Using 1-cm grid paper, make a bar graph to show the results of both surveys. According to your graph, what percentage of students liked jazz music? What percentage liked pop music? What percentage liked alternative music?

## Reflection

**1.** What did you know about the problem before you started to solve it?

**2.** What did you need to figure out?

**3.** How did you simplify or make a similar problem?

**4.** Do you think the "Solve a Simple or Similar Problem" strategy was a good strategy to use? Why?

## Extension

**1.** Working with a partner, survey the students in your class on a topic that interests you.

**2.** Record the results in a tally chart.

**3.** Now conduct the same survey among students in your school.

**4.** Share your results with another set of partners. How did the survey results for your classmates compare to the results for the students in the whole school?

## Problem 7

# Counting Eggs

STRATEGY: MAKE A MODEL

You can use different materials, such as egg cartons, construction paper, or craft sticks, to build a model for a problem. Using these materials helps you see a solution to the problem.

OBJECTIVE:

Look for a pattern

## Problem

You Will Need
- egg cartons or a picture of a rectangle divided into 12 parts
- counters

Li visited her grandparents' farm. She was given the task of collecting eggs each morning for a week. There are 10 hens on the farm.

1. If each hen can lay 1, 2, or 3 eggs every day, what is one possible pattern in which the hens might lay the eggs over 1 week?

2. Based on this pattern, what would be the total number of eggs laid in 1 week?

3. If your pattern continued for 3 weeks, how many dozens of eggs would have been laid at the end of that time?

**Tip**

Use counters and 7 sections of an egg carton to help you.

## Reflection

**1.** What did you know about the problem when you set out to solve it?

**2.** What did you need to figure out?

**3.** What do your models look like?

**4.** Do you think the "Make a Model" strategy was a good strategy to use? Why?

**5.** Share your work with your teacher.

## Extension

Make up your own problem about farm life for a classmate. See if he or she can use the "Make a Model" strategy to solve it.

## Problem 8

# Price of Stamps

STRATEGY: LOOK FOR A PATTERN

Looking for a pattern is an important strategy and it can be used for many kinds of problems. You can identify many different types of patterns, such as patterns in things you see, or in the numbers you are working with. You can then continue the pattern to find the solution to a problem.

OBJECTIVE:

Look for a pattern in stamp prices

## Problem

1. If the price of a stamp this year went up by $0.01 every 3 years, how much would a stamp cost in 2020? How much would a stamp cost in 2040? How much would a stamp cost in 2060?

2. What is the pattern?

## Reflection

1. What did you know about the problem when you set out to solve it?

2. What did you need to figure out?

3. How did you find the pattern?

4. Do you think the "Look for a Pattern" strategy was a good strategy to use? Why?

5. Is your solution to the problem reasonable? Explain.

## Extension

What would a stamp cost in the year 3000, if the price of a stamp this year increased by $0.01 every year?

## Problem 9

# Cell Phone Costs

*STRATEGY: CONSTRUCT A TABLE*

You can make a table to help you organize information, as well as to spot missing information. Constructing a table makes looking for patterns easier, and it also highlights important information to help you solve a problem.

*OBJECTIVE:*

Perform calculations to do with money

## Problem

Imagine that you have a cell phone. Each month, the cost for phone calls is $10.00 for the first hour, $20.00 for the second hour, and $30.00 for the third hour and every hour after that. When you get your first monthly bill, you want to check the amount of time you have used in 1 month. In week 1, you used 30 min. In week 2, you used 40 min. In week 3, you used 50 min. In week 4, you used 20 min. What would be your total charge for using the cell phone that month?

### Tip

To calculate the rate for the minutes used that don't make up a full hour, figure out the rate per minute based on the rate per hour.

## Reflection

1. What information did you have before you started to solve the problem?

2. What did you need to figure out?

3. Explain how you solved this problem.

4. When you set up your table, how many rows did you need?

5. Do you think the "Construct a Table" strategy was a good strategy to use? Why?

6. Compare your answer with a classmate's.

## Extension

The second month, you use the cell phone $\frac{3}{4}$ of the time you used it the first month.

1. For how much time did you use the cell phone in the second month?

2. For how much less time did you use it in the second month than the first month?

3. How much money did you save compared to the first month?

## Problem 10
# Number Combinations

STRATEGY: MAKE AN ORGANIZED LIST

Putting together a list allows you to organize your thinking and review what you have done. You can identify important steps in a problem, or things you need to figure out to help you solve the problem.

OBJECTIVE:

Recognize number patterns

## Problem

You are in charge of making three-digit race numbers to be worn by people competing in the town's bicycle race. The head official informs you that you can only use the numbers 4, 5, and 6, and each number can be used only once in each three-digit number. How many people can be in the race?

### Tip

You can use each digit only once in each three-digit number. For example, that means 444 and 556 are not possible.

## Reflection

1. What did you know about the problem as you set out to solve it?

2. What did you need to figure out?

3. How is your list organized?

4. Was the "Make an Organized List" strategy a good strategy to use? Why?

5. What pattern did you find?

6. Share your work with your teacher.

## Extension

What if you were given 5, 6, 7, and 8 and asked to make four-digit numbers? How many four-digit numbers could you make? (Remember, you can't repeat any of the digits in each four-digit number.)

# Unit 2
# Math and Flight

Since they first saw birds in flight, people have wondered what materials, shapes, and angles would allow them to glide and soar in the same way.

In Chapter 4, you will learn about different types of human-made fliers as you study two-dimensional geometric figures. You will then use what you have learned to build, test, and improve a trick flier that performs fractional loops, turns, and rolls.

In Chapter 5, you will focus on metric measurement. You will learn about the relationships among metric units, so that you can make conversions. At the end of the chapter, you will conduct a survey about where to locate an airport in your community and then make design plans for your airport.

In Chapter 6, you will consider important milestones in the history of flight and some aviation-related careers. Your math focus will be on three-dimensional geometry, volume and mass, and coordinate grids. At the end of the chapter, you will determine which three-dimensional shape will make the best wing.

**Chapter**

# 4

# The Math of Taking Off

Fractions and mixed numbers are needed to make flight-related measurements and calculations, and the 24-hour clock helps us keep track of flight arrivals and departures.

In this chapter, you will
- classify angles
- classify triangles
- classify quadrilaterals
- study symmetry in regular polygons
- compare and order fractions and mixed numbers
- explain the relationship between fractions and decimals
- describe the relationship between 12-hour and 24-hour clocks

At the end of this chapter you will design, build, test, and modify a paper trick flier that performs fractional loops, rolls, and turns.

Answer these questions to prepare for your adventure:

1. What makes shapes congruent or similar? How are these two terms the same? How are they different?

2. Explain how fractions and decimals are related.

3. What is a line of symmetry? What does a line of symmetry in a shape mean?

# Exploring Angles and Triangles

*FLIGHT PLAN*

*PLAN:*
*You will estimate, measure, and identify angles. You will also work with equilateral, isosceles, and scalene triangles.*

*DESCRIPTION:*
*You can see the angles involved in flight by watching jets take off at airports or perform at air shows. Triangles have different types of angles.*

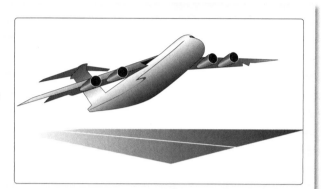

## Get Started

**1. a)** Where can you see angles in the picture in the Flight Plan?

  **b)** What math words could you use to describe each type of angle?

  **c)** How would you measure the angles?

### You Will Need
• protractor

**2.** Work with a partner. Use a protractor to measure each angle in the Flight Plan picture. Classify each angle according to the definitions in the vocabulary box.

### Journal

How did your results in question 1b) compare with your results in question 2?

### Vocabulary

**acute angle:** An angle that measures greater than 0° but less than 90°    45°

**obtuse angle:** An angle that measures greater than 90° but less than 180°    110°

**reflex angle:** An angle that measures between 180° and 360°    230°

**right angle:** An angle that measures exactly 90°    90°

**straight angle:** An angle that measures exactly 180°    180°

## Estimate, Measure, and Classify Triangles

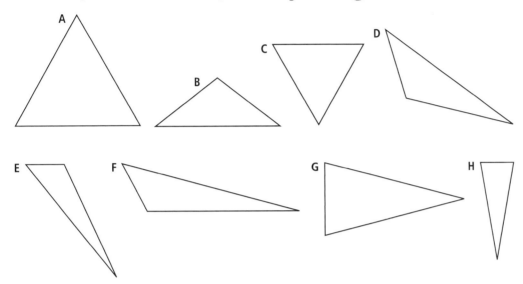

1. Without measuring, classify each triangle in the illustration as equilateral, isosceles, or scalene.

2. Measure the sides of each triangle to check your classification.

3. How did your estimates compare with the actual results?

4. Estimate what types of angles are in each triangle. Record your predictions.

5. Measure the angles of the triangles. Check whether your predictions were accurate.

### Vocabulary

**equilateral triangle:**
A triangle with all sides equal in length

**isosceles triangle:**
A triangle that has two sides of equal length

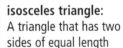

**scalene triangle:**
A triangle with three sides of different lengths

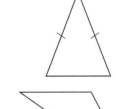

### Technology

Access a Web site that provides online explanations and practice on identifying angles and on identifying types of triangles based on their angles.
For example,
www.aaamath.com/g513-triangle-angles.html or
www.mathleague.com/help/geometry/angles.htm

1. Define in your own words acute, right, obtuse, straight, and reflex angles. Then define equilateral, isosceles, and scalene triangles. Give examples of different types of angles and triangles from the illustrations and photos in this book.

2. How did your ability to estimate help you to classify angles and triangles?

3. What did you learn about the measures of the angles in an equilateral, isosceles, and scalene triangle?

## Practice

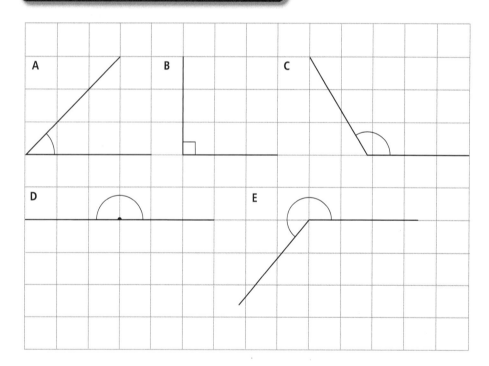

Complete the following steps for each angle:

1. Estimate the angle measurement and record your estimate.

2. Use a protractor to measure the angle.

3. Classify each angle as right, acute, obtuse, reflex, or straight. Give reasons to support your classification decisions.

4. In your notebook, use a ruler to draw examples of equilateral triangles, isosceles triangles, and scalene triangles. Label each triangle.

## Lesson 2

# Congruency and Similarity

### FLIGHT PLAN

PLAN:
You will learn how to identify and draw congruent and similar figures such as the ones in this hang glider.

DESCRIPTION:
The inventor Leonardo da Vinci dreamed of human flight. He observed birds and sketched plans for human-powered flying machines. Many years later, people created hang gliders that allowed them to soar like birds.

## Get Started

Congruent figures are the same size and shape, and their corresponding angle and side measures are equal.

Similar figures have the same shape but are not necessarily the same size. Their corresponding angles will be equal. Their corresponding sides will be proportional. If you multiply all of the sides of the smaller shape by the same number, you will get the side lengths of the larger shape.

You Will Need
• ruler
• protractor

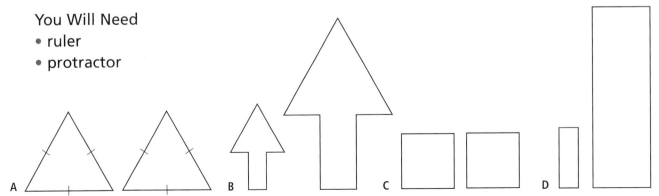

**1.** Which pairs of figures are congruent? Which pairs of figures are similar? How can you tell?

**2.** Prove that the figures are congruent or similar by measuring them. Record your results.

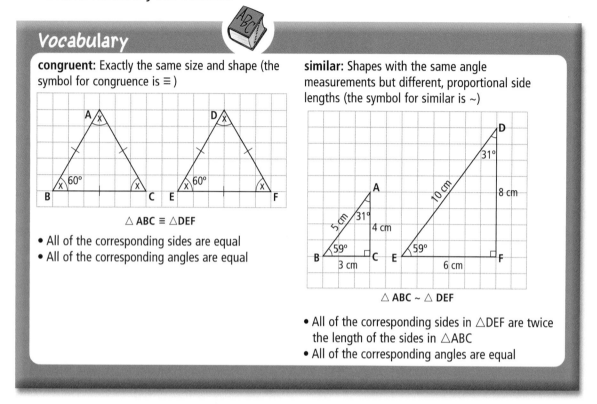

## Vocabulary

**congruent:** Exactly the same size and shape (the symbol for congruence is ≡)

△ ABC ≡ △DEF

- All of the corresponding sides are equal
- All of the corresponding angles are equal

**similar:** Shapes with the same angle measurements but different, proportional side lengths (the symbol for similar is ~)

△ ABC ~ △ DEF

- All of the corresponding sides in △DEF are twice the length of the sides in △ABC
- All of the corresponding angles are equal

## Build Your Understanding

## Identify and Create Congruent and Similar Figures

You Will Need
- ruler
- protractor

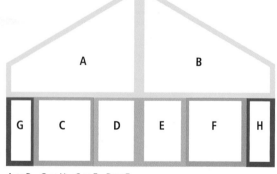

A ≡ B   G ≡ H   C ≡ F   D ≡ E

**1.** Work as part of a design team to create two-dimensional plans for a hang glider. Your plans must include sets of congruent and similar figures. Make sure that you can identify all congruent and similar figures in your design as well as explain why sets of figures are congruent or similar.

**2.** Exchange your plans with another design team. Challenge them to use a ruler and protractor to find all congruent and similar figures in your hang glider design.

## What Did You Learn?

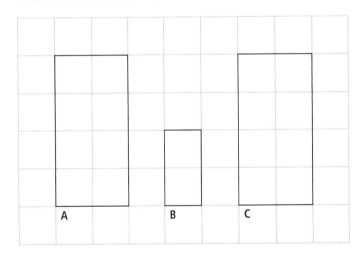

**1.** Which of the rectangles shown above are congruent and which are similar? How do you know?

**2.** What is the main difference between congruent figures and similar figures?

**3.** What measurement tools would you use to check for congruency? What measurement tools would you use to check for similarity?

You Will Need
• grid paper or a geoboard

1. Use grid paper or a geoboard and elastic bands to create pairs of congruent figures and pairs of similar figures. Make sure that you can explain why figures are congruent or similar.

2. Exchange your congruent and similar figures with a classmate and challenge him or her to identify which figures are congruent and which are similar, and explain why.

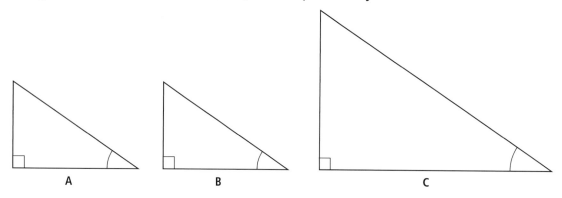

A        B                C

3. a) Estimate which figures shown above are congruent and which are similar. Record your choices.

   b) Use measurement tools to determine which figures shown above are congruent and which are similar.

4. Design a logo for a brand-new Canadian airline. Create a congruent version of the logo for a brochure and a version that is similar for a poster.

**Journal**

Write step-by-step instructions for determining whether two or more figures are congruent or similar.

**Technology**

Use a geometry application to create an online worksheet of various types of triangles based on the types of angles they have. Label each triangle. At the bottom of the screen, or using a word processor, create questions that a classmate will answer: for example, colour the scalene triangle blue; or, name all of the triangles that contain a right angle. Prepare an answer sheet and check your classmate's answers.

# Lesson 3

# Exploring Diagonals

## FLIGHT PLAN

PLAN:
You will explore and make generalizations about the diagonal properties of quadrilaterals such as those found in this kite.

DESCRIPTION:
Kites come in all sizes and shapes. They are important in many cultures. Kites were first made in China about 3000 years ago.

## Get Started

You Will Need
• protractor

The two centre pieces of a kite's frame are diagonals (a line segment that joins two vertices in a polygon that are not adjacent).

1. Estimate the angles where the diagonals intersect. Record your estimate.

2. Use a protractor to check your estimate.

### Journal

Sketch the shape of a kite you have flown or seen. Label the vertices. Draw diagonals on the shape. Indicate with an X where the diagonals intersect.

### Vocabulary

**bisect:** Divide into equal halves
**diagonal:** A line segment that joins two non-adjacent vertices across a two-dimensional shape

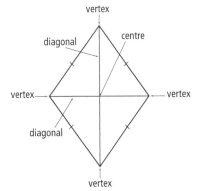

**intersect:** To have one or more points in common —intersecting lines cross each other at one point
**vertex:** The point where two rays of an angle or two edges meet in a plane figure, or the point where three or more sides meet in a solid figure (the plural of vertex is vertices)

## Explore Diagonal Properties of Quadrilateral Shapes

You Will Need
- protractor
- scrap paper
- scissors
- ruler

The quadrilateral is a common shape for kites. You might see the following quadrilateral shapes at a kite festival:

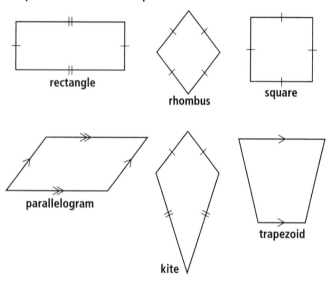

rectangle

rhombus

square

parallelogram

kite

trapezoid

**Vocabulary**

**kite:** A quadrilateral with two pairs of equal adjacent sides

**parallelogram:** A shape that has four sides with opposite sides that are parallel, and with opposite angles that are equal

**quadrilateral:** A polygon that has four sides

**rhombus:** A quadrilateral with four equal sides, and opposite angles that are equal

**trapezoid:** A quadrilateral with only one pair of parallel sides

**Tip**

Hatch lines indicate sides that are equal.

Parallel arrows indicate sides that are parallel.

(A side with single marks matches another side with single marks. A side with double marks matches another side with double marks.)

**1.** Copy this chart into your notebook.

| Quadrilateral | Do the Diagonals Bisect Each Other? | How Large Is the Angle Formed Where the Diagonals Intersect? |
|---|---|---|
| square | | |
| kite | | |
| rectangle | | |
| rhombus | | |
| parallelogram | | |
| trapezoid | | |

**2.** Draw a square, a kite, a rectangle, a rhombus, a parallelogram, and a trapezoid.

**3.** Cut out the quadrilaterals you drew and label them on the back as a square, kite, and so on.

**4.** For each quadrilateral, fold along all possible diagonals. Then draw in the diagonals.

**5.** Measure and record whether the diagonals for the quadrilateral bisect each other.

**6.** Use a protractor to measure the angles formed where any diagonals intersect. Record your measurements in your chart.

**Tip**

You will be folding the figures, so make them at least three or four times the size of the quadrilaterals shown in the illustration on page 164.

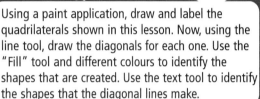

**Technology**

Using a paint application, draw and label the quadrilaterals shown in this lesson. Now, using the line tool, draw the diagonals for each one. Use the "Fill" tool and different colours to identify the shapes that are created. Use the text tool to identify the shapes that the diagonal lines make.

## What Did You Learn?

**1.** Compare and discuss your chart results with a classmate's. Recheck your results if they are different from your classmate's.

**2.** What patterns do you notice in your chart results?

**3.** What conclusions can you draw from the chart results?

## Practice

Draw the following angles. Use a protractor to help you.

**1.** 60°   **2.** 25°   **3.** 33°   **4.** 42°   **5.** 17°   **6.** 84°

Create the following shapes using some of the angles from questions 1 to 6:

**7.** equilateral triangle

**8.** parallelogram

**9.** isosceles triangle

**10.** Explain how you created the shapes in questions 7 to 9.

**11.** Work with a partner. One of you can say one clue about a certain type of triangle. For example, you might say, "I have three equal sides." Your partner must guess which triangle you are. Then switch roles.

**12.** An isosceles triangle has one angle that measures 62°. One of its other angles measures 56°. What might the third angle measure?

**13.** A scalene triangle has one angle that measures 102°. Another angle measures 45°. What will its third angle measure?

## Extension

**14.** Use a protractor to measure the angles at the vertices of each quadrilateral you explored in Build Your Understanding. For each quadrilateral, add all its angles. What pattern(s) do you notice in the sum of angles in the quadrilaterals?

**15.** Estimate the sum of angles in each triangle.

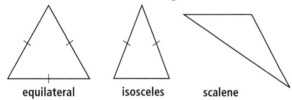

equilateral          isosceles          scalene

**16.** Use a protractor to measure the angles in the triangles above and check your estimate. What pattern(s) do you notice in the sum of angles in the triangles?

### Show What You Know

**Review: Lessons 1 to 3, Angles, Triangles, Congruency, and Diagonals**

**1.** Classify all the angles in this illustration:

**2.** In Lesson 2, you designed a logo for a brand-new Canadian airline. Look at your design. List all the congruent shapes and all the similar shapes you can find.

**3.** In your journal, explain the following terms in your own words: quadrilateral, parallelogram, kite, rhombus, trapezoid.

# Lesson 4

# Exploring Line Symmetry

FLIGHT PLAN

PLAN:
You will build on your understanding of symmetry and conduct experiments to find patterns in the number of lines of symmetry in regular polygons and quadrilaterals.

DESCRIPTION:
Both natural fliers, such as butterflies, and human-made fliers, such as kites, have lines of symmetry.

## Get Started

You Will Need
• tracing paper
• scissors
• ruler

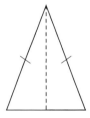

We say a figure has line symmetry if it can be divided into two congruent parts, each of which is a reflection of the other. This isosceles triangle can been folded along the dotted line so one half covers the other exactly.

The dotted line is called the line of symmetry. A line of symmetry divides a shape into two congruent parts.

### Journal

Make sketches of things in your classroom that have symmetry. Use a ruler to draw each figure's line or lines of symmetry. Label important parts of your diagrams.

### Vocabulary

**line of symmetry:** A line that divides a shape into two congruent symmetrical parts
**line symmetry:** Parts reflected in a line of symmetry
**regular polygon:** A polygon with all sides and angles equal
**symmetrical:** having parts that are balanced about a line or point

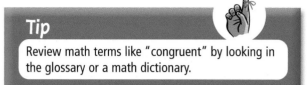

**Tip**

Review math terms like "congruent" by looking in the glossary or a math dictionary.

1. Work with a partner to identify which of the figures shown above have symmetry and which do not.

2. How can you test your conclusions?

3. For the figures you decided have symmetry, identify their line or lines of symmetry.

4. Some figures have more than one line of symmetry. The regular polygon shown here is an equilateral triangle.

   Use tracing paper to make a copy of the equilateral triangle. Cut out the triangle and fold it to find all possible lines of symmetry. Use a ruler and a pencil to draw in all the lines of symmetry you find. Share, compare, and discuss your findings with another pair of students.

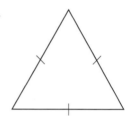

## Build Your Understanding

## Sort Regular Polygons by Lines of Symmetry

You Will Need
- regular polygons (equilateral triangles, squares, pentagons, hexagons, heptagons, octagons)
- scissors
- ruler

1. Copy and complete this chart for a square, a pentagon, a hexagon, and an octagon.

| Regular Polygons | How Many Lines of Symmetry? | Sketch the Lines of Symmetry |
|---|---|---|
| equilateral triangle | 3 | |

**2.** Work in a small group. Experiment with folding each regular polygon to find all its lines of symmetry. Draw the lines of symmetry along the folded lines using a pencil and a ruler.

**3.** Enter the results on your chart.

## What Did You Learn?

**1.** What patterns do you notice in your chart results?

**2.** What conclusions can you draw from these patterns?

**3.** Describe the relationship between the number of sides a regular polygon has and its number of lines of symmetry.

**4.** Draw and cut out another regular polygon. Find all its lines of symmetry. Does the relationship pattern hold? Explain.

## Practice

### Math Problems to Solve

**1.** Draw a two-dimensional shape that has two lines of symmetry.

**2.** Will the patterns you noted in What Did You Learn hold for quadrilaterals?

**a)** Copy and complete this chart in your notebook.

| Quadrilaterals | Predict the Lines of Symmetry | Sketch the Lines of Symmetry | What Patterns Did You Find? |
|---|---|---|---|
| rectangle | | | |
| rhombus | | | |
| parallelogram | | | |

**b)** Compare and discuss your findings with those of a classmate.

## Lesson 5

# Rotational Symmetry

*FLIGHT PLAN*

*PLAN:*
*You will learn about rotational symmetry and sort regular polygons according to their order of rotational symmetry.*

*DESCRIPTION:*
*Turning shapes such as propellers allow many flying machines to get off the ground.*

## Get Started

You Will Need
• ruler
• protractor
• tracing paper

Regular polygons, such as equilateral triangles, have rotational symmetry. This means that when you rotate or turn a tracing of the polygon around its centre, the tracing will fit exactly on the original polygon.

The order of rotational symmetry is the number of times a tracing fits on the original figure during one 360° rotation. The centre of rotation is the point around which the polygon is rotated.

In this diagram, the tracing was turned around the centre of rotation. The tracing fit exactly on the original equilateral triangle three times, so the order of rotational symmetry is three for this regular polygon.

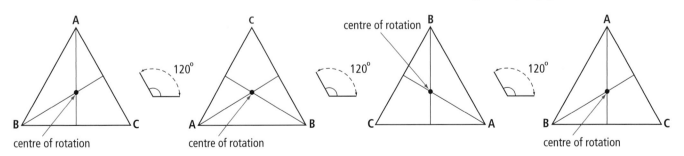

1. Use a ruler and a protractor to make a drawing of an equilateral triangle. Make sure the drawing outline is bold.

2. Trace the equilateral triangle.

3. To find the centre of rotation, draw a line from one vertex to the midpoint of the opposite edge, or to the opposite vertex (depending on the polygon). Then draw a line from a different vertex to the midpoint of its opposite edge, or to its opposite vertex. The point where these two lines intersect is the centre of rotation.

4. Turn the tracing on the original equilateral triangle around the centre of rotation to find the order of rotational symmetry.

## Journal

Write a definition of "rotational symmetry" in your own words. Draw and label a diagram of a polygon other than an equilateral triangle to support your definition.

## Vocabulary

**centre of rotation:** The point of intersection of two of the angle bisectors in a polygon or the point around which a polygon is turned

**order of rotational symmetry:** The number of times a tracing fits on the original figure during one 360° rotation around its centre of rotation

**rotation** or **turn:** A transformation (movement) of a figure around a fixed point (the figure does not change size or shape, but it does change position and orientation)

**rotational** or **turn symmetry:** When a shape can fit onto itself more than once during a full 360° turn

## Build Your Understanding

## Sort Regular Polygons by Rotational Symmetry

You Will Need
- regular polygons (listed in the chart on page 172)
- ruler
- protractor
- compass
- tracing paper

Work with a partner.

1. Draw each regular polygon from the chart on page 172. For each regular polygon, find the centre of rotation and label it with a dot on the original polygon.

2. Trace the outline of each regular polygon on tracing paper. Mark the centre of rotation on each tracing.

**3.** Place your tracing of the second regular polygon, the square, on the original square. Turn the tracing of the original square. Make note of how many times in a 360° turn the tracing fits exactly on the original square.

**4.** Record the order of rotational symmetry for each regular polygon in the chart. The equilateral triangle has been done for you.

**5.** Use this method to find the order of rotational symmetry for each regular polygon on the chart. Make sure to record your results.

| Regular Polygon | Order of Rotational Symmetry |
|---|---|
| equilateral triangle | 3 |
| square | |
| pentagon | |
| hexagon | |
| heptagon | |
| octagon | |
| nonagon | |
| decagon | |

**Journal**

Comment on any patterns you noticed as you found the order of rotational symmetry for the regular polygons.

**Tip**

You might use the point of a compass to hold the centre of rotation of the original shape and the tracing in place.

## What Did You Learn?

**1.** Explain to a classmate what "rotational symmetry" and "order of rotational symmetry" mean. Use manipulatives to make your explanations as clear as possible.

**2.** Meet with another pair to share, compare, and discuss your results. What patterns did you notice in your chart results? What conclusions can you draw from these results?

**3.** What relationship do you notice between the number of sides a regular polygon has and its order of rotational symmetry? Test your conclusions.

1. Estimate the size of each angle. Then, use a circular protractor to check your estimate. Identify whether each angle is acute, right, obtuse, straight, or reflex.

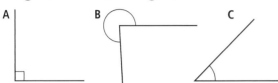

2. In your notebook, create a congruent figure for each polygon.

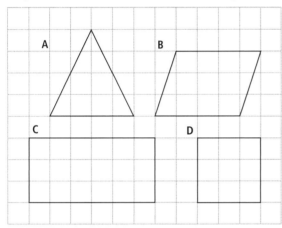

**Technology**

The helicopter propeller is one example of an object that has rotational symmetry. Use the Internet to find pictures of other objects that also show rotational symmetry. Paste these into a desktop publisher or paint application. Beside each picture, draw the polygon that is rotated.

3. Create a similar figure for each polygon.

4. Explain to a classmate why each figure you created is congruent or similar to the original.

**Journal**

Make note of figures that have rotational symmetry. Outline the steps needed to find the rotational symmetry of the figures you select.

5. Which of the letters shown are symmetrical? Write them in your notebook, showing their line or lines of symmetry.

# F H D A Z

## Show What You Know

**Review: Lessons 4 and 5, Symmetry**

1. Find all the lines of symmetry in this shape:

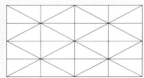

2. Which letters of the alphabet have lines of symmetry? What patterns do you notice?

**Lesson 6**

# Comparing Proper Fractions

*FLIGHT PLAN*

*PLAN:*

You will develop your understanding of proper fractions and learn how to compare and order these fractions on a number line.

*DESCRIPTION:*

This illustration shows the seating plan for a plane that carries 120 passengers in 14 first-class seats and 106 coach-class seats. People often use proper fractions to describe how many seats are filled on a flight. For example, $\frac{1}{2}$ of the seats are full for the flight.

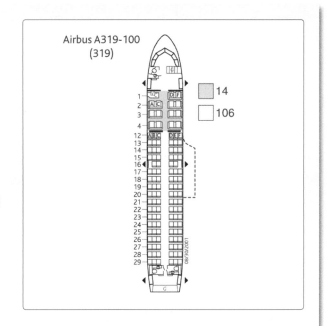

## Get Started

A fraction is a number that shows a part of a whole. In a proper fraction, the numerator is smaller than the denominator.

$$\frac{3}{4} \quad \begin{array}{l} \text{numerator} \\ \text{denominator} \end{array}$$

On one flight the first-class section of the plane is $\frac{5}{16}$ full and on another it is $\frac{1}{4}$ full.

**1.** Predict which fraction you think will be larger and record your prediction.

## Vocabulary

**denominator:** The bottom part of a fraction that tells how many parts are in the whole; for example, $\frac{3}{4}$

**equivalent fractions:** Fractions that name the same amount or part; for example, $\frac{1}{2}, \frac{2}{4}, \frac{3}{6}, \frac{4}{8}, \frac{5}{10}$, and $\frac{6}{12}$ are all equivalent fractions

**fraction:** A number that names a part of a whole; for example, $\frac{1}{4}$

**lowest terms:** A fraction or ratio in which both the numerator and the denominator have no common factor other than 1; for example, $\frac{1}{3}$ is in its lowest terms, while $\frac{2}{6}$ is not, even though both fractions represent the same amount or part

**numerator:** The top part of a fraction that tells how many parts are being referred to; for example, $\frac{3}{4}$

**proper fraction:** A fraction in which the numerator is smaller than the denominator; for example, $\frac{3}{4}$

**2.** Create two grids in your notebook like the ones shown below.

 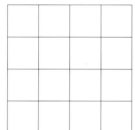

**3.** Colour the first grid to show that $\frac{1}{4}$ of the seats are occupied, and colour the second grid to show that $\frac{4}{16}$ of the seats are occupied. What do you notice about the two grids?

---

$\frac{4}{16}$ can be reduced to its lowest terms of $\frac{1}{4}$ because both fractions represent the same amount

$$\frac{4 \div 4}{16 \div 4} = \frac{1}{4}$$

When a fraction is reduced to its lowest terms, both the numerator and the denominator are divided by the same number, until neither the numerator nor denominator has a common factor other than 1.

For example: $\frac{2 \div 2}{4 \div 2} = \frac{1}{2}$

---

**4.** What patterns can you find when these fractions are reduced?
$$\frac{2}{12} = \frac{1}{6}; \quad \frac{2}{4} = \frac{1}{2}; \quad \frac{3}{9} = \frac{1}{3}$$
What rule can you state to reduce fractions to lowest terms?

**5.** Create two more grids like those in question 2. Colour the first grid to show that $\frac{1}{4}$ of the seats are occupied, and colour the second grid to show that $\frac{5}{16}$ of the seats are occupied. Which fraction is greater?

**6.** Use the symbols < (less than) and > (greater than) to write correct mathematical statements about the relationship between the two proper fractions in question 5.

## Compare and Order Proper Fractions

You Will Need
- scissors
- metre stick
- three strips of paper (7 cm x 21 cm)
- glue or tape

Work with a partner or in a small group.

1. Glue or tape the strips of paper together so that you have one strip that is 62 cm long.

2. Measure 1 cm from the left end of the strip, make a mark, and write "0" under it. At the right of the strip, 1 cm from the end, make a mark and write "1" under it. Use the metre stick to draw a line that connects the two marks.

3. Fold the strip in half lengthwise. Unfold the strip and mark $\frac{1}{2}$ at the centre.

4. Fold the strip into quarters lengthwise. Label the fold marks $\frac{1}{4}$, $\frac{2}{4}$, and $\frac{3}{4}$.

5. Fold the strip into eighths. Mark and label each eighth.

6. Use the metre stick to measure thirds, fifths, sixths, tenths, and twelfths.

   Fractions you labelled more than once such as $\frac{1}{2}$, $\frac{2}{4}$, $\frac{3}{6}$, and $\frac{5}{10}$ are called equivalent fractions. In lowest terms, these fractions are all $\frac{1}{2}$, and they represent the same part of the whole. What other equivalent fractions appear on your number line?

7. Use your number line to complete the following:

   **a)** $\frac{1}{3} = \frac{\blacksquare}{6}$ or $\frac{\blacksquare}{12}$  **b)** $\frac{2}{6} = \frac{\blacksquare}{12}$

   **c)** $\frac{2}{3} = \frac{\blacksquare}{6}$ or $\frac{\blacksquare}{12}$  **d)** $\frac{1}{2} = \frac{\blacksquare}{4}$ or $\frac{\blacksquare}{6}$ or $\frac{\blacksquare}{8}$ or $\frac{\blacksquare}{10}$ or $\frac{\blacksquare}{12}$

**8.** Estimate which is the greater fraction in each pair. Then, use the number line to check your estimates.

**a)** $\frac{2}{5}$ or $\frac{3}{8}$    **b)** $\frac{5}{8}$ or $\frac{3}{10}$

**c)** $\frac{5}{6}$ or $\frac{11}{12}$    **d)** $\frac{2}{3}$ or $\frac{3}{4}$

**9.** Use your number line to order these fractions from least to greatest:

**a)** $\frac{1}{4}, \frac{2}{10}, \frac{3}{8}, \frac{1}{6}, \frac{1}{3}$    **b)** $\frac{9}{10}, \frac{7}{8}, \frac{5}{6}, \frac{3}{4}, \frac{4}{5}$

**c)** $\frac{1}{2}, \frac{2}{3}, \frac{3}{10}, \frac{5}{8}, \frac{3}{8}$    **d)** $\frac{2}{3}, \frac{4}{5}, \frac{1}{4}, \frac{5}{8}, \frac{2}{10}$

**Tip**

Use skip counting—counting by twos, threes, fours, and so on—to help you identify equivalent fractions.

**Journal**

Outline the steps for comparing proper fractions. For example, suppose you wanted to compare $\frac{3}{16}$ and $\frac{1}{8}$. What steps would you follow? How could you apply this method when comparing and ordering other proper fractions?

## What Did You Learn?

**1.** Write a definition for proper fractions. Include a labelled example to support your definition.

**2.** Identify three equivalent fractions. Use grid paper, your number line, or other concrete materials to explain to a classmate why the fractions are equivalent.

**3.** Organize these fractions on a number line from least to greatest.

$\frac{1}{6}$    $\frac{1}{3}$    $\frac{2}{9}$    $\frac{5}{12}$    $\frac{1}{2}$    $\frac{7}{18}$

**Booklink**

**Gator Pie** by Louise Mathews and Jeni Bassett (Dodd, Mead: New York, NY, 1979). In this story, two alligators must divide their pie into many different fractions.

Which fraction is greater?

**1.** $\frac{3}{4}$ or $\frac{1}{2}$  **2.** $\frac{5}{6}$ or $\frac{7}{8}$  **3.** $\frac{4}{5}$ or $\frac{3}{8}$

For each fraction, find an equivalent fraction:

**4.** $\frac{2}{4}$  **5.** $\frac{5}{10}$  **6.** $\frac{1}{3}$

**7.** Order these fractions on a number line:

$\frac{3}{4}$  $\frac{2}{3}$  $\frac{1}{2}$  $\frac{7}{8}$

**8.** Which of the above fractions is greatest?

## Math Problems to Solve

**9.** If 14 of an airplane's seats are first class and 106 are coach class, what fraction of the plane's total number of seats are first class? Explain how you arrived at your answer. Use numbers, pictures, and words.

**10.** How many passengers are needed to make the plane $\frac{1}{2}$ full?

**Tip**

Remember that fractions are part of a whole. What is the whole in question 11?

**Journal**

Make note of anything you learned from solving the problems that you can use to improve your own problem-solving strategies.

# Mixed Numbers and Improper Fractions

*FLIGHT PLAN*

*PLAN:*

*You will compare and order mixed numbers and improper fractions on a number line.*

*DESCRIPTION:*

*Li and Claire went to an air show. Li paid 13 quarters, or $3.25, admission. After they purchased their tickets, they watched one of the aerobatic planes do a $1\frac{1}{2}$ roll. $1\frac{1}{2}$ is a mixed number and can also be written as $\frac{3}{2}$, which is an improper fraction.*

## Get Started

Li had 13 quarters. Since 4 quarters make a dollar, the amount Li had was $\frac{13}{4}$ dollars.

The number $\frac{13}{4}$ is an improper fraction, a fraction in which the numerator is greater than or equal to the denominator.

If we divide 13 by 4, we get the whole number 3 with a remainder of 1. This remainder is $\frac{1}{4}$ of the whole. The fraction $\frac{13}{4}$ can also be written as $3\frac{1}{4}$, which is a mixed number. A mixed number is a whole number and a fraction.

Work with a partner. One partner explains how this drawing illustrates the improper fraction $\frac{13}{4}$. Then, the other partner explains how the drawing shows the mixed number $3\frac{1}{4}$.

1. Work in pairs or a small group.

   **a)** Draw pictures in your notebook to show these fractions:
   $\frac{11}{4}$, $\frac{7}{4}$, $\frac{5}{4}$, $\frac{13}{12}$, $\frac{15}{12}$

   **b)** Draw pictures in your notebook to show these mixed numbers:
   $1\frac{1}{4}$, $2\frac{3}{4}$, $2\frac{1}{2}$, $1\frac{1}{2}$, $1\frac{3}{12}$

2. Draw pictures to show $3\frac{1}{2} > \frac{7}{4}$ and $\frac{5}{4} < 1\frac{1}{2}$. Order all four improper fractions and mixed numbers in this question from least to greatest on a number line.

**Vocabulary**

**improper fraction:** A fraction with a numerator that is larger than the denominator; for example, $\frac{17}{4}$
**mixed number:** A number that is part whole number and part proper fraction; for example, $7\frac{3}{4}$
**whole number:** The numbers belonging to the set [0,1,2,3, ...]

## Build Your Understanding

## Compare and Order Mixed Numbers and Improper Fractions

You Will Need
• paper
• metre stick
• scissors

1. Measure a number line that is 9 cm long. Cut out the number line.

2. Mark and label whole numbers from 0 to 8 on your number line as reference points. Each number represents 1 cm.

3. **a)** Order the following mixed numbers on your number line from least to greatest: $7\frac{1}{4}$, $5\frac{3}{8}$, $6\frac{1}{4}$, $4\frac{3}{8}$, $3\frac{5}{8}$, $2\frac{1}{2}$, $1\frac{7}{8}$. Mark where each mixed number is in relation to the nearest whole numbers.

   **b)** Write equivalent improper fractions for each mixed number.

**4. a)** On your number line, order from least to greatest the following improper fractions: $\frac{5}{4}$, $\frac{15}{2}$, $\frac{9}{8}$, $\frac{13}{4}$.

  **b)** Write equivalent fractions and mixed numbers for each improper fraction.

**Journal**

How could a calculator help you convert mixed numbers to improper fractions?

## What Did You Learn?

**1.** Explain to a classmate the strategy you used to convert an improper fraction to a mixed number, and to convert a mixed number to an improper fraction. Draw pictures to help make your explanation clear for the listener.

**2.** Give an example of an improper fraction that is slightly more than $3\frac{3}{4}$.

**3.** Give an example of a mixed number that is slightly less than $\frac{23}{7}$.

## Practice

Use your number line to help you place the correct symbols, < or >, between the following pairs of mixed numbers and improper fractions in your notebook. Try estimating first, then check your estimates.

**1.** $6\frac{3}{8}$ ▇ $\frac{24}{8}$      **2.** $\frac{17}{4}$ ▇ $2\frac{1}{2}$      **3.** $3\frac{1}{2}$ ▇ $\frac{8}{4}$

Draw pictures to show each mixed number or improper fraction:

**4.** $\frac{17}{4}$    **5.** $3\frac{2}{3}$    **6.** $4\frac{2}{5}$    **7.** $\frac{8}{6}$

Write each improper fraction as a mixed number:

**8.** $\frac{19}{5}$    **9.** $\frac{7}{4}$    **10.** $\frac{9}{5}$    **11.** $\frac{23}{8}$

Write each mixed number as an improper fraction:

**12.** $3\frac{2}{7}$    **13.** $2\frac{1}{3}$    **14.** $4\frac{3}{5}$    **15.** $2\frac{1}{6}$

**16.** In the improper fraction $\frac{5}{3}$, how many parts are in each whole? How many parts are there altogether? Which is the numerator? Which is the denominator? Show $\frac{5}{3}$ by drawing a picture or by using concrete materials. Write $\frac{5}{3}$ as a mixed number.

# Lesson 8

# Relating Fractions to Decimals

FLIGHT PLAN

PLAN:

To calculate the cost of a balloon trip, you will explore the relationship between fractions and decimals by making conversions.

DESCRIPTION:

For many years people have flown in balloons. Balloons are filled with hot air or gases that are lighter than air. For a fee, some tour companies will take passengers on balloon flights.

## Get Started

Sometimes it is useful to express fractions as decimals. For instance, Tony, his mother, and two friends want to take a balloon ride over their community. The cost of the balloon ride is $165.00. Since the children must raise their share of the fee, it is helpful to know exactly how much each person will pay in dollars and cents: $\frac{165}{4}$.

**1.** Use your calculator to find the fee for each person.

**2.** Use Method 1 to estimate and then convert $\frac{1}{5}$ to decimal form.

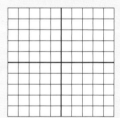

### Tip

A fraction represents division. To find the decimal equivalent of a fraction, you can divide the numerator by the denominator.

**Changing Proper Fractions to Decimals**

There are three useful ways to change fractions to decimals.

**Method 1: Using equivalent fractions**

You know that $\frac{45}{100} = 0.45$. What is $\frac{1}{4}$ as a decimal?

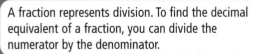

Think: What will I have to multiply 4 by to obtain a denominator of 100?

$\frac{1}{4} = \frac{1 \times 25}{4 \times 25}$

(if I multiply the denominator by 25, I must do the same to the numerator)

$= \frac{25}{100}$ (this is an equivalent fraction to $\frac{1}{4}$)

$= 0.25$

## Method 2: Using long division

You learned in question 1 that a fraction can represent division and that you can divide the numerator by the denominator to obtain a decimal.

Example, $\frac{1}{4}$:

$$
\begin{array}{r}
0.25 \\
4\overline{)1.00} \\
\underline{8}\phantom{00} \\
20 \\
\underline{20} \\
0
\end{array}
$$

## Method 3: Using a calculator

$\frac{1}{4} = 0.25$

$\boxed{1}\ \boxed{\div}\ \boxed{4} = 0.25$

---

**3.** Use Method 2 to write $\frac{3}{8}$ in decimal form.

**4.** Use Method 3 to find the decimal equivalent of $\frac{3}{7}$. Round to three decimal places.

## Changing Mixed Numbers to Decimals

When expressing a mixed number as a decimal, start with the fraction. For example, in $1\frac{7}{8}$, start with $\frac{7}{8}$. Think 7 divided by 8 and use Method 2 or 3.

$\frac{7}{8} = 0.875$

Then add the whole number to the decimal.

$1 + \frac{7}{8}$

$= 1.875$

## Changing Decimals to Fractions

When expressing a decimal number such as 2.255 as a fraction, first consider the digits to the right of the decimal point. These digits become the numerator, and the denominator becomes a multiple of ten. The number of zeroes in the denominator is the same as the number of digits right of the decimal point.

$0.255 = \frac{255}{1000}$

Express this as a fraction in lowest terms.

$\frac{255}{1000} = \frac{255 \div 5}{1000 \div 5} = \frac{51}{100}$

$2.255 = 2\frac{51}{100}$

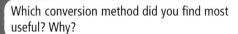

**Journal**

Which conversion method did you find most useful? Why?

---

**5.** Use Method 3 to express 3.95 as a fraction in lowest terms.

**6.** How could you use a calculator to check your work?

## Explore the Relationship Between Fractions and Decimals

**You Will Need**
- grid paper
- coloured pencils
- calculator

Use the strategies and tools you find most useful for expressing fractions as decimals or decimals as fractions. Check each answer using a calculator.

**Vocabulary**

**decimal:** A number based on 10 with one or more digits to the right of a decimal point; for example, 1.03

**rounding:** A rule used to make a number approximate to a certain place value. Numbers are rounded up when the digit immediately following the required place value is 5 or higher, and rounded down when the digit is less than 5. For example, 238 rounded to the nearest 10 is 240. 1238 rounded to the nearest 100 is 1200.

**1.** Write each fraction as a decimal.

   **a)** $\frac{3}{4}$    **b)** $\frac{4}{5}$    **c)** $\frac{1}{8}$    **d)** $\frac{13}{20}$

**2.** Use grid paper to show how you converted one fraction to a decimal. With a classmate, share, compare, and discuss the conversion strategy you applied.

**3.** Write the following mixed numbers and improper fractions as decimal numbers. If necessary, round to three decimal places.

   **a)** $2\frac{1}{4}$    **b)** $5\frac{1}{2}$    **c)** $4\frac{3}{5}$

   **d)** $3\frac{1}{3}$    **e)** $\frac{11}{4}$

**Tip**

If you were expressing $\frac{2}{3}$ as a decimal and dividing 2 by 3 using a calculator, you would see 0.6666666..., a repeating decimal. To round this repeating decimal to three decimal places, look at the fourth number after the decimal place, which is 6. Since 6 is more than 5, round to 7. Rounded to the third number after the decimal place, $\frac{2}{3} = 0.667$.

**4.** Express each decimal as a fraction.

   **a)** 1.5    **b)** 0.11    **c)** 0.75    **d)** 1.775    **e)** 4.135

**5.** Convert each fraction pair to decimals. If necessary, round your answer. Then, determine which is greater. Indicate which is least or greatest using the symbols < or >.

   **a)** $\frac{5}{8}$ ▦ $\frac{7}{9}$    **b)** $\frac{8}{9}$ ▦ $\frac{4}{5}$    **c)** $\frac{1}{3}$ ▦ $\frac{2}{5}$

   **d)** $\frac{7}{8}$ ▦ $\frac{3}{5}$    **e)** $\frac{23}{8}$ ▦ $\frac{226}{76}$

**6.** Arrange all decimal numbers in question 5 on a number line from least to greatest.

1. Choose a method to express $\frac{1}{20}$ in decimal form. Explain why you chose this method. Draw a picture on grid paper to show how you arrived at your answer.

2. Write $\frac{5}{6}$ in decimal form. Round to three decimal places. Check your answer with your calculator.

3. Outline the steps you would follow to find the decimal equivalent of $1\frac{3}{7}$.

4. Express 1.775 as a fraction. Make sure that you write it in its lowest terms.

**Practice**

Express the following as decimals. Round to three decimal places. Find one of the answers using two different methods.

1. $\frac{14}{6}$     2. $\frac{5}{12}$     3. $3\frac{1}{3}$     4. $\frac{3}{10}$

Express the following as fractions and reduce to their lowest terms if necessary:

5. 0.236     6. 0.34     7. 0.16     8. 0.9

9. 0.124     10. 0.52     11. 0.25     12. 0.8

13. In the balloon question in Get Started, another friend wants to join the original group of four. How much will each group member now have to pay? Express your answer as a decimal.

14. What is the next fraction in this sequence of fractions: $\frac{1}{9}, \frac{2}{9}, \frac{3}{9}, ...$? Express all four fractions as decimals using a calculator. What do you notice?

## Lesson 9

# Reading the 24-Hour Clock

**FLIGHT PLAN**

**PLAN:**
You will learn about the relationship between 12-hour and 24-hour clocks.

**DESCRIPTION:**
Keeping track of time using a 24-hour clock is important at any Canadian or international airport.

| Time | | Origin | Vol | | Remarks |
|------|------|--------|-----|------|---------|
| | Expected | | Flight | | News |
| 0930 | 0929 | Amsterdam | NW | 8927 | |
| 0930 | | New York | SR | 111 | Delayed |
| 0930 | | New York | DL | 111 | |
| 0940 | 0912 | Marseille | LX | 731 | |
| 0955 | | Zurich | SR | 934 | |
| 0955 | | Zurich | DL | 2640 | |
| 1000 | | Francfort | LH | 3678 | |
| 1000 | | Francfort | UA | 3694 | |
| 1000 | | Geneve | SO | 3821 | Cancelled |
| 1000 | 0947 | Madrid | SR | 3653 | |

## Get Started

On the 24-hour clock, all A.M. numbers appear in red. All P.M. numbers are in blue. Instead of the hours running from 1 to 12, the hours on the 24-hour clock run from 1 to 24. Notice that times on a 24-hour clock have 4 digits (01:00 is 1:00 A.M.).

1. Work with a partner to convert the following times to the 24-hour clock.

   **a)** 3:00 P.M.    **b)** 4:00 A.M.

   **c)** 4:00 P.M.    **d)** 8:00 P.M.

   Discuss your answers and conversion strategies with another pair of students.

### Vocabulary

**A.M.:** Before noon; the time between midnight and noon
**digits:** Numerals from 1 to 9
**P.M.:** After noon; the time between noon and midnight
**24-hour clock:** A clock based on 24 hours, starting from 00:00 through to 24:00 hours

**2.** Create questions on a 12-hour clock for a classmate to convert to a 24-hour clock. Make sure that you know the answer to each of your questions so you can check your classmate's answers.

## Build Your Understanding

### Conversions Between 12-Hour and 24-Hour Clocks

You Will Need
- construction paper
- white paper
- scissors
- compass or protractor
- black, red, and blue coloured pencils

**1.** Work with a partner. Use the tools and materials to make a 24-hour clock with a 12-hour clock in the centre, like the one shown in Get Started. Outside the clock face, mark A.M. times in red and P.M. times in blue.

**2.** Test each other's understanding of the 24-hour clock by using the clock you made to ask and answer conversion questions.

**3.** Copy the following chart in your notebook and complete it using the clock you made.

| 24-hour clock | 12-hour clock A.M. | 24-hour clock | 12-hour clock P.M. |
| --- | --- | --- | --- |
| | 1:00 A.M. | 13:00 | 1:00 P.M. |
| 05:00 | | | 2:00 P.M. |
| | 7:00 A.M. | 18:00 | |
| 09:00 | 9:00 A.M. | 20:00 | 8:00 P.M. |
| | 11:00 A.M. | 22:00 | |
| 03:00 | | 24:00 | |

## What Did You Learn?

**1.** What are the easiest times on a 24-hour clock to read and convert? Why?

**2.** What are the most difficult times on 24-hour clock to read and convert? Why? What strategies might you use to help you make these conversions?

**3.** Look closely at your conversion chart data. What patterns do you notice among the numbers?

**4.** Meet in a small group to share and discuss any relationships you noticed between times on the 12-hour and 24-hour clocks.

## Practice

Write these times as they would appear on a 24-hour clock:

**1.** 9:00 A.M.      **2.** 10:25 P.M.      **3.** 4:12 P.M.

**4.** 3:15 A.M.      **5.** 8:45 P.M.      **6.** 9:50 P.M.

**7.** 11:30 P.M.     **8.** 10:05 A.M.

Write these times as they would appear on a 12-hour clock:

**9.** 02:00      **10.** 24:00      **11.** 05:00

**12.** 16:25     **13.** 12:32     **14.** 17:00

**15.** 20:20     **16.** 15:45

### Extension

**17.** Eric lives in Hamilton, ON, and he is going to visit his grandmother in Victoria, BC. His flight leaves at 19:00 and takes 4 hours and 55 minutes. What time should Eric tell his grandmother to meet him at the airport according to the 12-hour clock and Victoria time?

**Tip**

Victoria, BC, is in a different time zone from Hamilton, ON. You might need a time-zone map of Canada to solve and create conversion problems.

## Show What You Know

**1. a)** Mahalia went to the movies with her family. She used 33 quarters to buy one ticket. Express this amount as an improper fraction.

**b)** Convert your improper fraction to a mixed number.

**c)** What would the fraction be if Mahalia used 33 dimes? What would it be if she used nickels?

**2.** Ajay, Li, Eric, Ruth, and Ahanu were playing a game. Out of 10 games, Li won $\frac{1}{5}$ games, Ajay won 0.1 games, Eric won 0.3 games, Ahanu won $\frac{1}{5}$ games, and Ruth won 0.2 games. How many games did each student win? Show your work using numbers, words, and pictures. What patterns do you notice?

**3.** Use a 24-hour clock to help you solve this problem. It takes 3 hours to fly from Calgary to Toronto. It takes 3 hours to fly from Toronto to Mexico. If you leave Calgary at 07:00 and you stop in Toronto for 1 hour, what time would it be in Mexico when you landed? (Remember that there is a time difference between Calgary and Toronto, and also between Toronto and Mexico. Use the Internet to help you find these time differences.)

# Chapter Review

1. **a)** Identify the types of angles shown in the illustrations below and estimate the measurement of each angle.

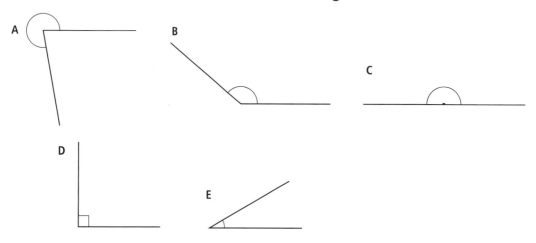

    **b)** Now, measure each angle using a protractor. How accurate were your estimates?

    **c)** Draw and name an acute angle, a right angle, and a reflex angle. Label each angle measurement.

    **d)** Draw each of the following angles: 35°, 140°, 80°.

2. **a)** Identify the types of triangles shown. Estimate the length of each side and each angle measurement.

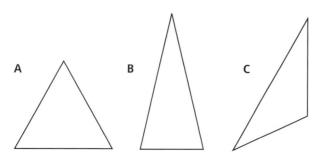

    **b)** Now, measure the angles and sides of each triangle. How accurate were your estimates? How might you improve them?

    **c)** Draw an equilateral triangle, an isosceles triangle, and a scalene triangle. Label the angle and side measurements for each.

**3.** Which of the shapes shown below are congruent and which are similar? How do you know?

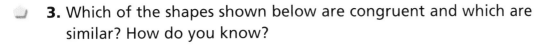

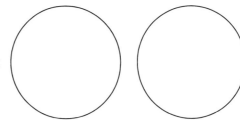
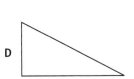

A

B

C

D

E

**4.** Draw a regular pentagon. Create angle bisectors and mark the centre of rotation. Make a tracing of your pentagon. What is the order of rotational symmetry?

**5. a)** Draw a regular polygon and a quadrilateral of your choice.

   **b)** Name the two shapes you have drawn.

   **c)** Sketch all lines of symmetry for each shape.

**6.** Order the following proper fractions, improper fractions, and mixed numbers from least to greatest:

   **a)** $\frac{3}{8}$, $\frac{1}{4}$, $\frac{14}{4}$, $2\frac{1}{2}$

   **b)** $3\frac{1}{5}$, $\frac{12}{10}$, $\frac{7}{5}$, $\frac{4}{5}$

   **c)** $\frac{7}{3}$, $2\frac{1}{6}$, $1\frac{1}{2}$, $\frac{2}{3}$

**7.** Convert the following fractions to decimal numbers. If the decimal number is repeating, round to three decimal places.

a) $\frac{3}{6}$    b) $\frac{5}{8}$    c) $\frac{11}{13}$    d) $\frac{17}{20}$

**8.** Convert the following decimals to fractions. If possible, reduce to their lowest terms.

a) 0.25    b) 0.45    c) 0.04    d) 0.50    e) 0.455

**9.** What would the following 12-hour clock times be on a 24-hour clock?

a) 2:00 P.M.          b) 3:35 A.M.

c) 6:00 P.M.          d) 8:45 A.M.

**10.** What would the following 24-hour clock times be on a 12-hour clock?

a) 12:00    b) 23:00    c) 08:00    d) 05:00    e) 18:00

**11.** An airplane leaves Montréal for Ottawa at 14:00. This flight takes a $\frac{1}{2}$ hour. The pilot continues on to Toronto where the plane lands after 90 minutes.

a) How long did the entire flight take?

b) What time did the plane land in Toronto?

# Chapter Wrap-Up

You have reached the end of Chapter 4. Throughout the chapter you have learned about flight, while classifying angles, constructing similar and congruent figures, sorting quadrilaterals according to their diagonals, comparing fractions and mixed numbers on a number line, explaining the relationship between fractions and decimals, and describing the relationship between 12-hour and 24-hour clocks. Your task now is to apply this knowledge and construct a paper flier which will perform stunts such as loops, turns, and rolls.

**You Will Need**
• project log
• rectangular pieces of paper
• scissors

## The Trick Flier

**1. a)** Fold the top of a piece of paper to one side.

**b)** Unfold it and do the same thing to the other side. Unfold. You now have a big X on your paper.

**c)** Fold the X in half from top to bottom. Push the sides in to the centre. The paper now looks like a rectangle with a triangle on top.

**d)** Fold the top of the triangle to the bottom of the triangle and press it flat.

**e)** Fold airplane in half from side to side. All folds should be on the inside. Create a fold about 2 cm in from the folded edge. Now flip the flier over and do the same thing on the other side.

**f)** Fold the wings out. Fold the edges of the wings up. Cut a flap in the back of each wing. These flaps are called rudders. You can fold the flaps up or down to control the airplane.

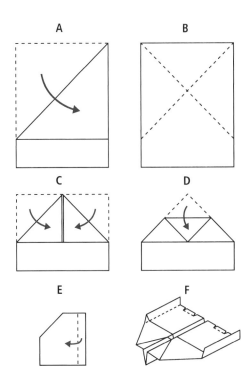

2. Work in a small team to build a flier. Test and compare the flight performance of your fliers.

3. Keep a project log for each step as you build and fly your plane. As you construct the flier, make note of the following:
   - acute, obtuse, right, straight, or reflex angles
   - shapes that are similar or congruent (show that they are similar or congruent)
   - shapes that are symmetrical (show that these shapes are symmetrical)
   - shapes that have rotational symmetry (show that the shapes have rotational symmetry)

4. Record the times in your log entries according to 24-hour clock time. Record equivalent 12-hour clock times beside the 24-hour clock times.

5. Adjust the flaps on the flier's wings or the tail rudders so that it performs turns, loops, and rolls.

6. In your project log, describe the turns, loops, and rolls. For example, "My flier did a $\frac{1}{2}$ turn," or "When we adjusted the rudders, the flier performed a $\frac{3}{4}$ turn." Be creative and use your flier to perform amazing stunts.

7. Display your completed project flight log on a classroom bulletin board.

**Journal**

Repeat steps A to F to make another flier using different sized paper. Repeat steps 2 to 5. What similarities and differences do you notice between your two fliers? Record them in your journal.

**Technology**

Use a Web site that contains instructions for making your own airplane. Create a second flier that is different from the one you made in this lesson. Compare the two airplanes using the instructions listed in the Chapter Wrap-Up section. Compare how well the two fliers perform when tested, and suggest reasons for the differences in performance.

# Chapter 5

# Math at the Airport

In this chapter, you will

- use appropriate metric units to measure dimensions and distance
- use formulas to calculate the perimeters and areas of different polygons
- find ratios and solve ratio problems
- determine the probability of different events occurring
- design surveys and collect, organize, and analyze data

At the end of this chapter you will conduct a survey about where to locate a new airport in your community and create design plans for that airport.

Answer these questions to prepare for your adventure:

1. How do you find the perimeter and area of a shape?
2. List the relationships among mm, cm, dm, m, and km.
3. What are three things that you need to do when creating and conducting a survey?

**Lesson 1**

# Metric Units of Measurement

*FLIGHT PLAN*

*PLAN:*
You will review and learn about different metric units used to make linear measurements. Then you will choose and use appropriate metric units to measure dimensions and distances around your school.

*DESCRIPTION:*
Linear metric measurements are used to describe the length, width, and height of aircraft, and how far they can fly. These measurements let you compare the sizes and flight capabilities of planes.

## Get Started

The metric measurement system is based on units of 10. One metre is the basic unit of measurement.

If one metre is divided into 10 equal parts, each part is $\frac{1}{10}$ of a metre, or a decimetre (the prefix deci- means "tenth"). There are 10 decimetres in 1 metre.

If each decimetre is divided into 10 equal parts, each part is $\frac{1}{100}$ of a metre, or a centimetre (the prefix centi- means "hundredth"). There are 100 centimetres in 1 metre.

If each centimetre is divided into 10 equal parts, each part is $\frac{1}{1000}$ of a metre, or a millimetre (the prefix milli- means "thousandth"). There are 1000 millimetres in 1 metre.

## You Will Need
- metre stick
- scissors
- scrap paper
- adhesive tape

1. With a partner use a metre stick and scissors to measure and cut out pieces of paper that are each a metre, decimetre, centimetre, and millimetre long.

2. Which length was the most difficult to create? Why?

3. With your partner, identify items in your classroom that you estimate are about one metre long, one decimetre long, one centimetre long, and one millimetre long.

4. Use a ruler to check your estimates.

5. One kilometre is 1000 metres. Identify a distance you regularly travel that is about one kilometre.

6. What measurement tools could you use to check your estimate of 1 km?

**Journal**

To help you understand metric units and make better estimates, record the dimensions or distances of familiar things or destinations—for example, the height of a doorknob from the floor is about one metre.

## Build Your Understanding

### Compare Units of Metric Measurement

You Will Need
- metre stick
- calculator

**Vocabulary**

**centimetre (cm):** $\frac{1}{100}$ of one metre

**decimetre (dm):** $\frac{1}{10}$ of one metre

**kilometre (km):** 1000 metres

**metre (m):** A unit of measurement used to measure length

**millimetre (mm):** $\frac{1}{1000}$ of one metre

Ajay learns that Plane A has the following dimensions.

| Plane A | | | | |
| --- | --- | --- | --- | --- |
| Wingspan | Length | Height | Maximum Altitude | Range |
| 11.1 m | 9.75 m | 2.8 m | 13 105 m | 1002 km |

He learns that Plane B has a height of 285 cm. He converts the height of Plane A to centimetres so that he can compare the two aircraft. To convert from a larger to a smaller unit, you must multiply. In 1 m there are 100 cm, so to convert from metres to centimetres, multiply by 100.

2.8 x 100 = 280

Plane A is 280 cm in height.

Ajay knows that the maximum height that Plane A can comfortably fly is 13 105 m. This is called the altitude. He wants to know what this altitude is in kilometres. To convert from metres to kilometres is to change from a smaller unit to a larger unit. When converting from a smaller to a larger unit, you must divide. There are 1000 m in 1 km, so to convert from metres to kilometres, divide by 1000.

13 105 ÷ 1000 = 13.105

Plane A can fly at an altitude of 13.105 km.

**1.** Use your calculator to check Ajay's calculations.

**2.** Copy the chart shown below into your notebook.

| Kilometres (km) | Metres (m) | Decimetres (dm) | Centimetres (cm) | Millimetres (mm) |
|---|---|---|---|---|
| 1 | 1000 | | | |
| | 1 | 10 | 100 | 1000 |
| | | 1 | | |
| | | | 1 | |
| | | | | 1 |

**3.** Use a metre stick, strips of paper, and the conversion strategies you learned in this lesson to complete the chart.

**4.** Use a calculator to check the accuracy of each conversion calculation.

**5.** What patterns do you notice?

**6.** Use the chart to help you convert the following measurements into metres. Write parts of a metre as decimals and then as mixed numbers and improper fractions.

   **a)** 326 cm      **b)** 115 cm

   **c)** 816 cm      **d)** 2 m 15 cm

   **e)** 8 m 56 cm

**7.** Arrange the measurements shown above in order from smallest to largest on a number line.

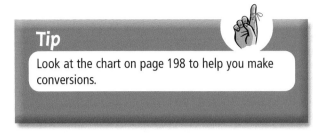

**Tip**

Look at the chart on page 198 to help you make conversions.

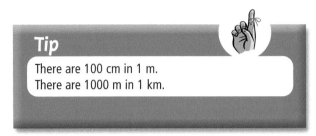

**Tip**

There are 100 cm in 1 m.
There are 1000 m in 1 km.

## What Did You Learn?

**1.** Describe in numbers, pictures, and words the relationships between the following:

   **a)** a millimetre and a centimetre

   **b)** a millimetre and a metre

   **c)** a decimetre and a centimetre

   **d)** a metre and a kilometre

**2.** Explain to a classmate the steps and operations you would follow to convert:

   **a)** a measurement from a larger to a smaller metric unit

   **b)** a measurement from a smaller to a larger metric unit

   Use examples to make your explanations as clear as possible.

**3.** What does the prefix tell you about a metric measurement unit's relationship to a metre?

## Practice

Use what you know about the relationships among metric units to complete the following statements.

   **1.** 3 dm = ▇ cm or ▇ mm

   **2.** 2 m = ▇ dm or ▇ cm or ▇ mm

   **3.** 6 cm = ▇ mm

   **4.** 3.5 km = ▇ m

Change each measurement to the unit in brackets.

**5.** 56 cm (dm)  **6.** 35 cm (mm)

**7.** 8 dm (mm)  **8.** 11 000 m (km)

What would be the most appropriate unit for measuring the following objects' dimensions? Why?

**9.** the length of a pencil

**10.** the width of a pencil

**11.** the thickness of a dime

**12.** the distance across your province

Which metric units and measurement tools would you use to measure the following? Give reasons to support each of your measurement decisions.

**13.** the distance between two cities in the same province

**14.** the width of a passenger seat on an airplane

**15.** the length of a jet's wing

**16.** Estimate and record the measurement of each item listed above. Measure or research the actual length, distance, width, or thickness. Comment on how the actual measurement compared with your estimate.

## A Math Problem to Solve

**17.** Two planes are waiting for takeoff on an airport runway. Plane A has a wingspan of 2590 cm and Plane B has a wingspan of 644 dm. Which plane has the larger wingspan? Outline the steps you will follow to solve the problem.

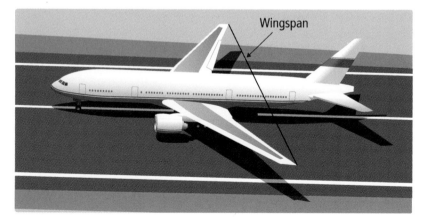

## Extension

You Will Need
- metre stick
- measuring tape such as one used for sewing
- construction measuring tape
- trundle wheel

Work in a small group.

**18. a)** Choose ten items with linear dimensions or distances to measure.

   **b)** In your notebook, copy and complete a chart similar to this one.

| Object | Measurement Unit | Measurement Tool | Estimate | | | Measurement | | |
|---|---|---|---|---|---|---|---|---|
| | | | Length | Width | Height | Length | Width | Height |
| | | | | | | | | |
| | | | | | | | | |

**Lesson 2**

# Finding Perimeters

FLIGHT PLAN

PLAN:
You will explore patterns in the dimensions and perimeters of polygons.

DESCRIPTION:
As this aerial photograph shows, an airport contains many different shapes. Sometimes it may be necessary to find the distance around these shapes, such as when fences need to be built.

## Get Started

Perimeter is the distance around an object. What is the perimeter of each of these shapes? How did you find the perimeter?

**1.**

**2.**

**3.**

4 cm
8 cm

**4.**

4 cm    1 cm
3 cm

---

**Journal**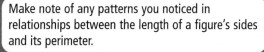

Make note of any patterns you noticed in relationships between the length of a figure's sides and its perimeter.

**Vocabulary**

**irregular polygon:** A polygon with side and angle measurements that are not equal
**perimeter:** The distance around an object

---

## Build Your Understanding

### Explore Perimeter Patterns

You Will Need
- 1-cm grid paper
- 30 square tiles (1 cm x 1 cm)

**1.** Find the perimeter of each square. Explain how you found the perimeter. What patterns do you notice?

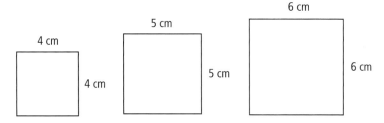

4 cm

5 cm

6 cm

4 cm

5 cm

6 cm

**2.** Imagine that another square has a side length of 7 cm. Predict what its perimeter will be. Use your tiles to make a square with these dimensions and check your prediction.

**3. a)** Use square tiles to make several different rectangles.

   **b)** Make a chart and record the lengths, widths, and perimeters of each rectangle.

**4.** Opposite sides of a parallelogram are parallel and equal. Calculate the perimeter of this parallelogram. Explain your calculation.

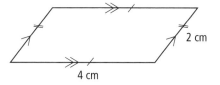

2 cm

4 cm

## What Did You Learn?

**1.** Describe the relationship between a rectangle's sides and its perimeter.

**2.** What relationship do you notice between an equilateral triangle's sides and its perimeter? Draw two equilateral triangles with different dimensions and calculate their perimeters to see whether the pattern holds.

**3.** What relationship do you notice between a parallelogram's sides and its perimeter? Draw two parallelograms with different dimensions and calculate their perimeters to check whether the pattern holds.

### Journal

Write step-by-step instructions explaining the easiest way to find the perimeter of a square, rectangle, equilateral triangle, and parallelogram.

Find the perimeters of the following shapes:

**1.** a square with side lengths of 2 cm

**2.** a rectangle 3 cm long and 200 cm wide

**3.** a parallelogram 1 m long and 200 cm wide

**4.** a square with side lengths of 5 m

**5.** an equilateral triangle with side lengths of 4 dm

**6.** an isosceles triangle with side A measuring 30 cm, side B measuring 30 cm, and side C measuring 15 cm

## Math Problems to Solve

**7.** Pearson International Airport in Toronto has 4 main runways and 30 taxiways. The runways have these dimensions:

- Runway 1 measures 3368 x 60.9 m
- Runway 2 measures 3300 x 60.9 m
- Runway 3 measures 2895 x 60.9 m
- Runway 4 measures 2590 x 60.9 m

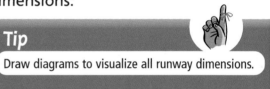
### Tip
Draw diagrams to visualize all runway dimensions.

Find the perimeter of each runway. Use a calculator to help you.

**8.** Look at the figure below.

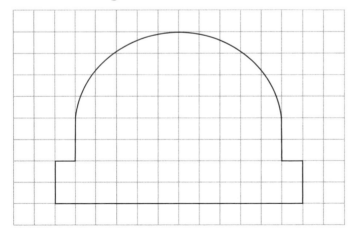

**a)** Estimate the perimeter of this figure and record your estimate.

**b)** What is the best strategy for measuring the perimeter of this irregular figure?

**c)** What is the best metric unit for measuring the perimeter of a figure of this size?

**d)** Measure the figure and calculate its perimeter.

**9.** Explain to a classmate the strategies you used to solve questions 7 and 8.

**Technology**

Use a spreadsheet application with cells formatted to a square shape (for example, column width and row height = 1 cm or 15 pt). Create squares, rectangles, and parallelograms of different sizes by filling the cells with colour. (You might want to use the copy and paste feature to fill cells more quickly.) If each cell represents 1 cm, calculate the perimeter of your polygons. Write an equation for each.

## Lesson 3

# Calculating Perimeter

*FLIGHT PLAN*

*PLAN:*
**You will learn and apply formulas for finding the perimeters of squares, rectangles, triangles, and parallelograms.**

*DESCRIPTION:*
Jets and other aircraft need runways of a certain size to take off and land. When planning airports, architects, engineers, and builders must make sure runways have the correct dimensions and perimeters.

## Get Started

Claire knows that an airport development is planned on a section of land that is 5 km by 7 km. Claire wants to build a safety fence, so she needs to find the perimeter of this piece of land.

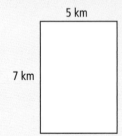

5 km

7 km

She knows one way to find the perimeter is to add all the side lengths:

7 + 5 + 7 + 5 = 24

Perimeter = length + width + length + width

Claire looks for patterns in lengths of the sides and discovers that

Perimeter = 2 x 7 + 2 x 5

Perimeter = 2 x length + 2 x width

$P = 2l + 2w$, or $P = 2(l + w)$

You Will Need
• 1-cm grid paper

**1.** Use the formula for perimeter ($P = 2l + 2w$) to check the perimeter of the rectangle of land. Draw several rectangles with different dimensions on 1-cm grid paper. Add the lengths and widths to find the perimeters. Then apply the formula to calculate the rectangle's perimeter. Which method of determining perimeter do you find easiest? Why?

A storage building at the airport has sides that are each 75 m long. Claire applies the formula $P = 2l + 2w$. She remembers that all sides of a square are equal, so she can simplify the formula even further to $P = 2s + 2s$, or $P = 4s$.

To solve her perimeter problem, Claire substitutes the dimension 75 for the variable $s$.

$P = 4 \times 75$

$P = 300$

The perimeter of the square building is 300 m.

**2.** This warning marker is an equilateral triangle. Estimate the perimeter, then record your estimate. Find the triangle's perimeter by adding side lengths.

**3.** Suggest a formula for finding the perimeter of equilateral triangles. Use the formula to check the perimeter of the warning marker. Draw several equilateral triangles. Add the side lengths to find the perimeters, then apply the formula. Which method of determining perimeter do you find easier? Why?

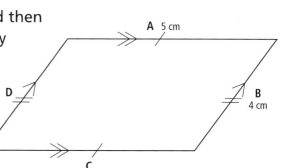

**4.** Estimate the parallelogram's perimeter and then record your estimate. Find the perimeter by adding the side lengths. What patterns do you notice in the side lengths? Suggest a formula for finding the perimeter of a parallelogram. Use your formula to find the perimeter.

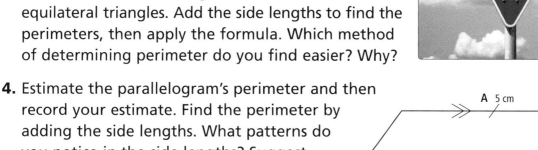

## Apply Formulas to Find Perimeters

You Will Need
- ruler
- calculator
- 1-cm grid paper

**1.** Work with a partner. For which figures can you use formulas to find the perimeters? Why?

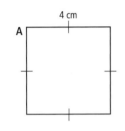

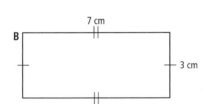

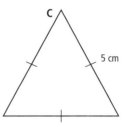

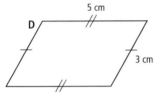

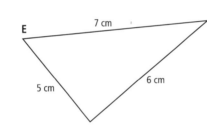

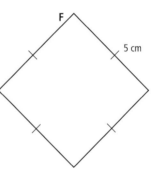

**2.** Estimate and record the perimeter of each polygon.

**3.** Find the perimeter of the polygon.

**4.** Check your answers.

### Vocabulary

**formula:** An open sentence that expresses a general rule; for example, the following are formulas for finding the perimeters of common polygons:
   **rectangle:** $P = 2l + 2w$
   **square:** $P = 4s$
   **equilateral triangle:** $P = 3s$
   **parallelogram:** $P = 2l + 2w$

**rhombus:** A quadrilateral with four equal sides, and opposite angles that are equal

**variable:** A letter or symbol that is used in an equation to represent an unknown or missing value; for example, in $2l + 2w$, $l$ and $w$ are variables representing length and width

## What Did You Learn?

**1.** With a partner, discuss your results and the formulas you used to calculate the perimeter.

**Tip**

You can draw a polygon and use it as an example in your explanation.

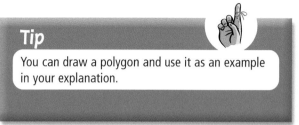

**2.** On poster paper, write the formulas for finding the perimeter of a rectangle, square, equilateral triangle, and parallelogram. Work in a group of four. Each group member selects a formula, then explains what the formula variables mean and how the formula can be applied.

## Practice

Find the perimeter of these shapes. State the formula used for each.

**1.** a rectangle (length = 21 cm, width = 13 cm)

**2.** a square (one side = 4 m)

**3.** a equilateral triangle (one side = 14 dm)

**4.** a parallelogram (length = 200 mm, width = 160 mm)

**5.** a rhombus (one side = 52 cm)

List the perimeter formulas that you would you use for these triangles.

**6.** an equilateral triangle

**7.** an isosceles triangle

**8.** a scalene triangle

Use linking cubes or tiles to show these perimeters:

**9.** a square (one side = 4 cubes)

**10.** a rectangle (length = 6 cubes, width = 4 cubes)

**11.** Use linking cubes or tiles to make different perimeters. Record the perimeter for each shape, and the formula you used. Challenge a partner to determine the perimeters you made. What formula did your partner use? Check that your partner's perimeter measurements are correct.

## Extension

**12.** Which strategies or formulas would you use to find the perimeters of these triangles? Explain why.

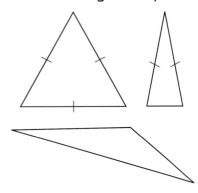

**13.** Suggest a formula for finding the perimeter of a rhombus. Use the formula to find the perimeter of this rhombus.

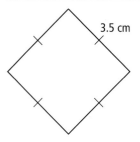

3.5 cm

## Show What You Know

### Review: Lessons 1 to 3, Measurement and Perimeter

**1.** Ajay's dad has a driveway that he wants to frame with pieces of wood. The wood comes in pieces that are 6 m, 8 m, and 10 m. Use the diagram below to determine what the best combination of wood pieces would be for Ajay's dad to frame the driveway.

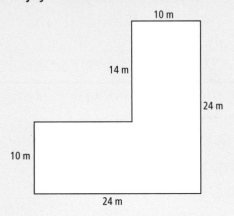

10 m
14 m
24 m
10 m
24 m

**2.** Work with a partner and measure 10 different objects in your classroom. Use a chart to record the objects and their measurements. What measurement units did you use? What patterns did you find?

## Lesson 4

# Finding the Areas of Polygons

FLIGHT PLAN

PLAN:
You will explore and develop rules for finding areas of rectangles and squares. As well, you will learn about units for measuring area.

DESCRIPTION:
People who build airports need to know the size of the surfaces they must cover. Airport runways are made of concrete, asphalt, or a combination of both materials. Tiles and carpeting are used to cover inside regions at airports.

## Get Started

Area is the amount of surface inside a two-dimensional shape, such as the rectangle that makes up a runway.

The area of a surface is measured in square units such as square millimetres ($mm^2$); square centimetres ($cm^2$); square decimetres ($dm^2$); square metres ($m^2$); square kilometres ($km^2$).

### Vocabulary

**area:** the amount of space inside a two-dimensional shape; area is measured in square units: $mm^2$, $cm^2$, $dm^2$, $m^2$, $km^2$

1. This square has an area of 1 $cm^2$. What is the perimeter of the square? How do the measurement units differ between perimeter and area?

1 cm

1 cm

2. How can you find the area of this rectangle? The rectangle is made up of three 1 cm x 1 cm squares.

## Explore Area

You Will Need
• tiles (30)
• grid paper

1. Imagine that each tile is 1 cm². Use your tiles to create squares with the following areas, or draw them on grid paper:

    **a)** 4 cm²    **b)** 9 cm²    **c)** 16 cm²

2. What pattern(s) do you notice in the relationship between the side lengths and the areas of squares?

3. Suggest a formula for finding the area of a square based on this pattern.

4. **a)** Copy a chart like this one into your notebook.

| Rectangle # | Length | Width | Rectangle Area |
|---|---|---|---|
|  |  |  |  |

**b)** Make as many rectangles as you can using 24 tiles or squares on grid paper in each rectangle. For each rectangle you make, record its length, width, and area in the chart.

**c)** What pattern(s) do you notice in the relationships between the lengths and widths and areas of the rectangles you created?

5. Suggest a formula for finding the area of a rectangle based on the pattern.

**Tip**
Use your calculator to check all area calculations.

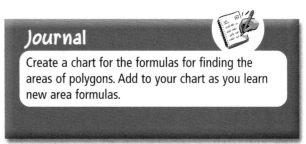

**Journal**
Create a chart for the formulas for finding the areas of polygons. Add to your chart as you learn new area formulas.

## What Did You Learn?

1. Work with a partner and take turns explaining the strategies you would use to find the area of a square. Then explain the strategy you would use to find the area of a rectangle.

2. Discuss in a small group:
   a) What relationship did you notice between the area of a square and its side lengths?
   b) What relationship did you notice between the area of a rectangle and its length and width?

3. In this lesson, you learned about the following units for measuring area: $mm^2$, $cm^2$, $dm^2$, $m^2$, and $km^2$. Which measurement unit would you use to measure the area of the following? Explain why.
   a) a runway       b) your province
   c) a napkin       d) one computer keyboard key

## Practice

1. Work with a partner to estimate the area of these squares.

2. Find the area of each square by counting or using any formula you discovered in the lesson.

### Technology

Using a paint or draw application draw a square that is 1 cm x 1 cm. Copy and paste this square to create various sizes of rectangles. Beside each rectangle, complete:
$l =$
$w =$
area =
See how many different rectangles you can create with an area of 30 $cm^2$. Use the formula for area to verify that these rectangles actually have an area of 30 $cm^2$.

2 cm

4 cm

6 cm

**3.** Estimate the area of each rectangle.

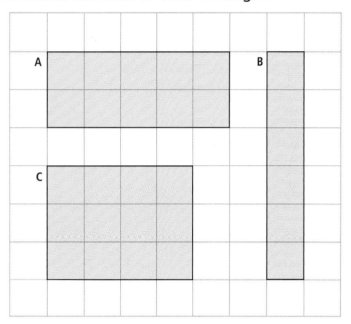

**4.** Find the area of each rectangle above by counting or using any formula you discovered in the lesson.

Use a geoboard to construct the following:

**5.** a square with an area of 25 cm$^2$

**6.** a square with a side length of 4 cm; then find the area

**7.** two different rectangles, each with an area of 16 cm$^2$

**8.** a rectangle with an area of 12 cm$^2$

Use linking cubes or tiles to show these areas:

**9.** a square with an area of 16 cm$^2$

**10.** a rectangle with an area of 24 cm$^2$

Solve the following problems:

**11.** If a square has an area of 25 cm$^2$, what must one of its side lengths measure? Explain your answer.

**12.** If a rectangle has an area of 18 cm$^2$, what might its length and width measure? Explain your answer.

## Lesson 5

# Factors and Area

*FLIGHT PLAN*

*PLAN:*
You will explore the relationships among the dimensions and area of a rectangle based on factors. Then you will develop and apply strategies to solve area problems.

*DESCRIPTION:*
From the window seat of an aircraft, you can see vast sections of urban and rural land once you are high in the air. How might you determine the areas of these large regions to compare their sizes?

## Get Started

You have discovered that the formula for the area of a rectangle is
Area = length x width
$A = l \times w$

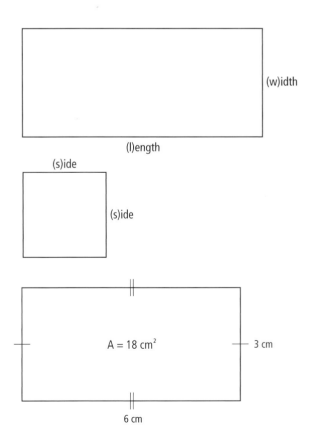

You have also found that the formula for the area of a square is
Area = side x side
$A = s \times s$

The length and width of this rectangle are factors of 18. A factor is a number that divides evenly into another number.

## You Will Need
• tiles (30)

**1.** For each number of tiles, list all the pairs of factors that, when multiplied together, give that number as their product. Record these factors in the second column. Extend the chart to 20 tiles.

| Number of Tiles | Factors of the Number of Tiles |
|---|---|
| 12 | 1 x 12,  2 x 6,  3 x 4 |
| 13 | |
| | |

**2.** Use tiles to make rectangles with lengths and widths equal to the factors in each of the above sets. What is the area of each rectangle?

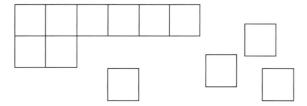

### Vocabulary

**factor:** A whole number that is multiplied by another whole number to find a product; for example, 3 is a factor of 6 because 3 x 2 = 6
**product:** The answer to a multiplication question

### Journal

Describe any patterns you notice among the factors, their products, and the areas of the rectangles.

## Build Your Understanding

### Solve Area Problems

How might you use your knowledge of area formulas to find the area of this large piece of land? There are two strategies for finding the area. One involves adding and the other requires subtraction.

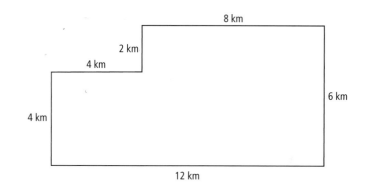

**Adding**

Since you know the formulas for finding the areas of rectangles and squares, you can add areas of the two parts of the large section of land.

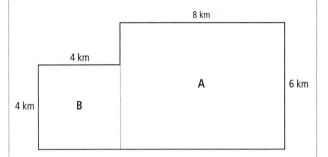

Area of A

$A = l \times w$

$= 6 \times 8$

$= 48$

Area of B

$A = l \times w$,
since B is a square
$A = s \times s$

$+ \quad = 4 \times 4$

$+ \quad = 16$

The area of A is 48 km².

The area of B is 16 km².

$48 + 16 = 64$

The total area of the section of land is 64 km².

---

**Subtracting**

You can also solve the problem by finding the area of the large rectangle and then subtracting the area of the small rectangle that is not part of the L-shaped section of land.

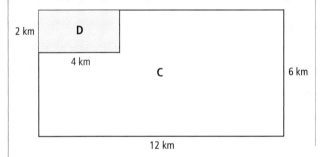

Area of C

$A = l \times w$

$= 6 \times 12$

$= 72$

Area of D

$A = l \times w$

$- \quad = 4 \times 2$

$- \quad = 8$

The area of C is 72 km².

The area of D is 8 km².

$72 - 8 = 64$

The total area of the section of land is 64 km².

---

1. Which of the two methods of finding area do you prefer? Why?

2. Work with a partner. Estimate the area of this figure and record your estimate.

3. Which strategy would you use to find the area of this figure? Outline the steps you will follow to solve the problem. Identify formulas you will apply to find the entire area.

4. Find the area of the figure.

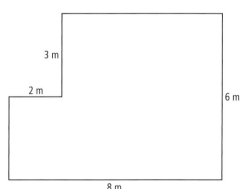

## What Did You Learn?

**1.** What are the pairs of factors of 36?

**2.** Write a statement explaining how the lengths, widths, and areas of rectangles are related to factors and their products. Draw a picture to support your explanation.

**3.** Meet with a small group of classmates and discuss these questions:

a) What strategy could you use to find the entire area of this figure?

b) What formula or formulas would you need to apply?

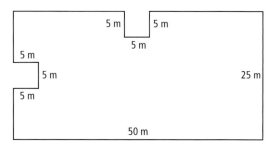

**4.** Use the most effective strategy and relevant formula(s) to find the area.

## Practice

Find the areas of these shapes and record the formulas used.

**1.** a rectangle 30 mm long and 5 cm wide

**2.** a square (one side = 22 cm)

**3.** a rectangle 3 m long and 50 dm wide

**4.** a square (one side = 10.2 cm)

**5.** a rectangle 30.9 m long and 62.3 m wide

> **Tip**
> Remember to convert measurements so that you are working with the same units.

**6.** If the area of a square is 25 cm², what are the factors that represent the dimensions?

**7.** If the area of a rectangle is 48 cm², what might the factors be that represent the dimensions? List three possible combinations.

## A Math Problem to Solve

**8.** From the air, you see a building like the one below, and you want to find the surface area of the flat roof.

a) Estimate the area and record it.

b) Identify a strategy or strategies for finding the area.

c) Identify the formula or formulas you will need to apply. Find the area.

d) Sketch a picture showing how you will solve the problem.

**Lesson 6**

# Parallelogram and Triangle Areas

*FLIGHT PLAN*

*PLAN:*
*You will explore the relationships among the areas of rectangles, triangles, and parallelograms.*

*DESCRIPTION:*
*At an airport you can find many different polygons, such as parallelograms and triangles.*

## Get Started

A parallelogram is a four-sided polygon with opposite sides that are parallel and equal. How is it similar to and different from a rectangle? The important parts of a parallelogram are its base and its height. Let's explore the relationships between a rectangle and a parallelogram.

You Will Need
• scissors
• grid paper
• adhesive tape

**1.** Draw a rectangle on a piece of grid paper. Estimate the area of your rectangle and record the estimate. Then, measure the rectangle and use the correct formula to find its area.

**2.** Make a line on your rectangle like the one shown to the right. Make a fold on the line and cut along the fold. Move the triangle to the opposite end of the rectangle to create a parallelogram (see illustration). Tape the triangle to the original figure.

**3.** What is the area of the parallelogram you created? How can you tell?

**Tip**

Did the length of the base stay the same? Did the height stay the same?

**4.** Draw a rectangle on grid paper, then cut it out. Draw a diagonal to divide the rectangle in half. What polygon did you create on each side of the diagonal?

**5.** Predict the area of one of the triangles by counting full and partial centimetre squares.

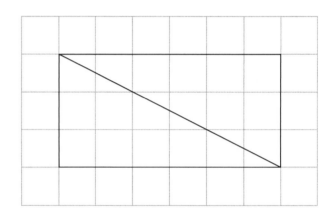

**6.** Find and record the area of the original rectangle. Now find the actual area of the triangle. How accurate was your prediction?

**7.** Write a statement to summarize the relationship between the area of a rectangle and the area of a triangle.

**Vocabulary**

**base** (of a polygon): A side of a polygon or a face of a solid figure by which the figure is measured or named; the symbol for base is $b$

**height:** A line perpendicular to the base through the top vertex; the symbol for height is $h$

**Booklink**

**Factors and Multiples** by William Fitzgerald (Addison-Wesley: Menlo Park, CA, 1986). This book will help you practise your factors and multiples.

## Find Patterns in Areas

You Will Need
• geoboard
• elastic bands

**1.** Make a chart like this one in your notebook.

| Rectangle Length | Rectangle Width | Rectangle Area | Parallelogram Base | Parallelogram Height | Parallelogram Area |
|---|---|---|---|---|---|
|  |  |  |  |  |  |

**2.** Work with a partner. On the geoboard create a rectangle. Determine the length and width of the rectangle and its area. Record this information in the appropriate chart cells.

**3.** Create a parallelogram by moving the top two vertices of the rectangle one space to the left. Determine the base and height of the parallelogram and its area. Record this information in the appropriate chart cells.

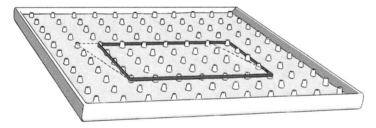

**4.** Create a chart like this one in your notebook.

| Rectangle Length | Rectangle Width | Rectangle Area | Triangle Area |
|---|---|---|---|
|  |  |  |  |

**5.** On the geoboard create a rectangle. Determine the length and width of the rectangle and its area. Record this information in the appropriate chart cells.

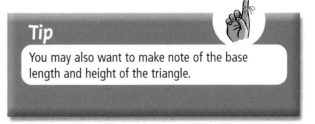

**Tip**

You may also want to make note of the base length and height of the triangle.

**6.** Use an elastic band to create a
diagonal that divides the rectangle
into two congruent triangles.
Determine the area of one triangle
and record it in the appropriate
chart cell.

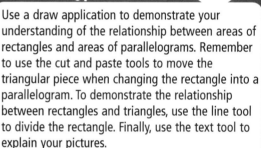

## What Did You Learn?

**1.** Look at your results in the first chart.
What relationship did you notice
between the areas of rectangles
and the areas of parallelograms
made from those rectangles?

**2.** Look at your results in the second
chart. What relationship do you
notice between the areas of
rectangles and the areas of triangles
made from those rectangles?

### Technology

Use a draw application to demonstrate your
understanding of the relationship between areas of
rectangles and areas of parallelograms. Remember
to use the cut and paste tools to move the
triangular piece when changing the rectangle into a
parallelogram. To demonstrate the relationship
between rectangles and triangles, use the line tool
to divide the rectangle. Finally, use the text tool to
explain your pictures.

## Practice

Sketch the following:

**1.** a square with an area of 9 cm$^2$

**2.** a rectangle with an area of 15 cm$^2$

**3.** a figure that is a combination of a square with an area of 4 cm$^2$
and a rectangle with an area of 12 cm$^2$

Answer the following questions:

**4.** On 1-cm grid paper, draw five different rectangles. Label the
length and width, and calculate the area of each rectangle.

**5.** Use a diagonal to divide each rectangle you created into two
equal triangles. Choose one triangle in each rectangle, and
identify and label its base and height. Now calculate its area.

**6.** Draw five different parallelograms. Label the base and height,
and calculate the area of each parallelogram.

## Lesson 7

# Using Formulas for Area

FLIGHT PLAN

PLAN:
You will learn the formulas for finding the areas of triangles and parallelograms, and you will apply these formulas to check your estimates.

DESCRIPTION:
How many parallelograms and triangles can you spot in this illustration? If you were to estimate the area of each polygon, how could you check the reasonableness of your estimate?

## Get Started

The formula for the area of a parallelogram is

Area = base x height

$A = b \times h$

$\quad = 3 \times 2$

$\quad = 6$

The area of this parallelogram is 6 cm².

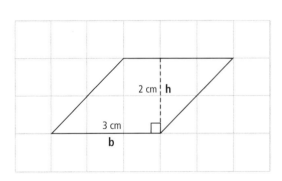

**1.** Use the formula to find the area of this parallelogram.

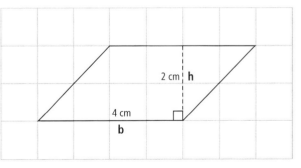

The formula for the area of a triangle is

Area = base x height ÷ 2

$A = b \times h \div 2$

$\quad = 3 \times 2 \div 2$

$\quad = 3$

The area of the triangle is 3 cm².

**2.** How is the area of the parallelogram related to the area of the triangle?

**3.** Estimate the area of this triangle and record your estimate. Then, use the formula to find the area. Check your answer by counting and adding full and partial squares.

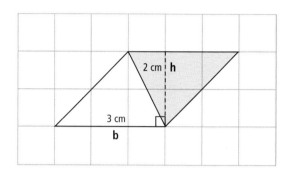

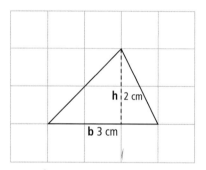

## Build Your Understanding

### Estimate and Calculate Areas

You Will Need
• grid paper
• ruler
• calculator

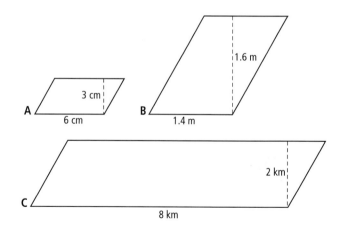

**1. a)** Identify the base and the height of each parallelogram.

  **b)** Estimate the area of each parallelogram.

  **c)** Calculate the area of each parallelogram. How reasonable were your estimates?

**2.** Copy and complete this chart in your notebook. Choose the most appropriate measurement unit in which to express the area and make any necessary conversions.

| Base | Height | Area of Parallelogram |
|---|---|---|
| 8 cm | 3 cm | |
| 3.2 m | 1.5 m | |
| 1500 m | 4.5 km | |

**3. a)** Identify the base and the height of each triangle shown below.

   **b)** Estimate the area of each triangle.

   **c)** Calculate the area of each triangle. How reasonable was your estimate in each case?

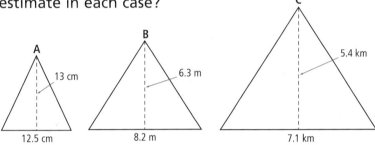

A    13 cm    12.5 cm

B    6.3 m    8.2 m

C    5.4 km    7.1 km

**4.** Copy and complete this chart in your notebook.

| Base | Height | Area of Triangle |
|---|---|---|
| 6.1 mm | 4.2 mm | |
| 4100 cm | 2.5 m | |
| 8.3 km | 2.1 km | |

**5.** Draw the parallelograms described in the chart. Draw a diagonal in each parallelogram to create two triangles. Shade one triangle in each parallelogram. Estimate the area of each shaded triangle. Use the correct formula to calculate area. How reasonable was your estimate?

Explain to a classmate the steps you would follow to find the area of the figures shown below. Assume each square on the grids is 1 cm². What area formulas could you apply? What operations would you need to perform? Use this strategy to find the area.

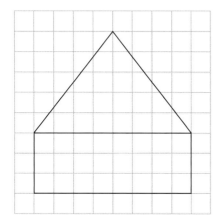
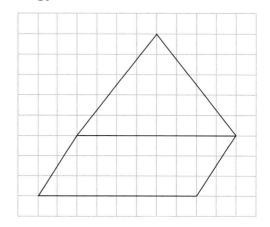

**Practice**

Calculate the area of each parallelogram. State the formula you used.

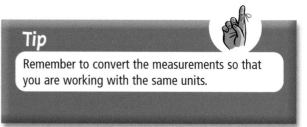
**Tip**
Remember to convert the measurements so that you are working with the same units.

1. base = 6.2 m, height = 7.3 m

2. base = 3.4 m, height = 60 cm

3. base = 200 cm, height = 100 cm

4. base = 49 dm, height = 20 m

5. Draw a parallelogram on grid paper with an area of 24 cm². What is its height measurement? What is its base measurement?

6. Draw a parallelogram on grid paper with an area of 36 cm². Divide the rectangle into two triangles. What is the area of one triangle? What formulas did you use?

## Math Problems to Solve

**7.** Work with a classmate.

    **a)** Estimate the area of this figure.

    **b)** Outline a strategy for finding the area of the figure.

    **c)** Use the strategy to calculate the area.

    **d)** Share, compare, and discuss your solution and problem-solving strategy with another pair of classmates.

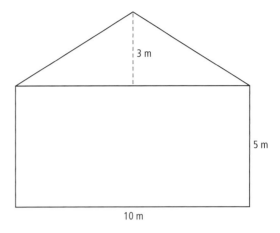

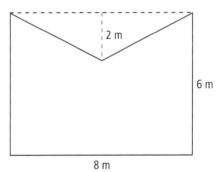

**8. a)** Airport workers are covering a portion of the floor with carpet in one of the rooms in an airport. Each roll of carpet will cover about 14 m². How many rolls of carpet will the airport workers need?

    **b)** How much would they pay for the carpet if each roll costs $112.00?

**9.** How did you solve these problems? Compare the strategies you used with a classmate's.

### Tip

What information do you know? What information do you need to know to solve the problem? What operations do you need to perform? What formulas can you apply? Make a step-by-step plan to solve the problem. Then, follow the plan to find a solution. You will need to round up your answer to the nearest full roll of carpet.

### Technology

You can use your knowledge of the formula for finding the area of a triangle in a spreadsheet, too. Launch a spreadsheet application and type the column headings: Triangle #, Base, Height, and Area of Triangle. In Row 2, input the number 1 in the first column, then type the base measurement in the second column and the height measurement in the third column. To calculate the area, click in the cell in the next column. In the formula entry window at the top of the screen, remember to type an = first. Remember also that to multiply you use the asterisk (*) and the dividing sign is / . Once you have completed the formula, press the <Enter> key and look at your answer. Does it make sense based on what you have learned about finding the area of a rectangle?

## Show What You Know

A candy company needs different boxes for their candy. Each box has a different price.

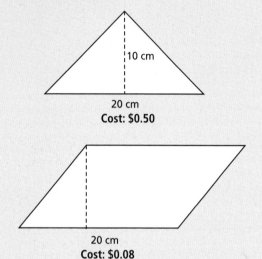

10 cm

20 cm
**Cost: $0.50**

20 cm
**Cost: $0.08**

**1.** Find the area of each box top.

**2.** Based on the top of each box, which box would hold the most candy?

**3.** If the company was going to buy 1000 of the boxes that hold the most candy, how much would the company spend?

**4.** If the chocolates had a length of 3 cm and a width of 1 cm, how many would fit in across the top of each box?

**Lesson 8**

# Exploring Ratio

*FLIGHT PLAN*

*PLAN:*
*You will learn to write ratios to express comparisons. As well, you will solve simple ratio problems.*

*DESCRIPTION:*
*The Concorde is a supersonic jet (supersonic means that it travels faster than the speed of sound). A typical Concorde flight from London to New York takes about three and a half hours (210 min). On regular airplane flights, the trip can take about eight hours (480 min). How can you compare the time it takes the two different types of aircraft to complete the flight?*

## Get Started

A ratio allows you to compare quantities that have the same units. When comparing the travel times of the two jets using ratios, you can use minutes or hours as common units.

The ratio of the number of minutes a Concorde flight takes to the number of minutes a subsonic flight takes is 210 : 480 (subsonic means that it travels slower than the speed of sound). The numbers 210 and 480 are the terms of the ratio 210 : 480. The order in which a ratio's terms are written is very important. By changing the order of ratio terms in the first sentence, you also change the ratio's meaning.

**210 : 480**

minutes the Concorde takes / minutes the subsonic aircraft takes

**480 : 210**

minutes the subsonic aircraft takes / minutes the Concorde takes

Ratios can be written in three ways.

- The number of minutes for a Concorde flight compared with the number of minutes for a subsonic flight is 210 to 480.

- The number of minutes for a Concorde flight : number of minutes for a subsonic flight is 210 : 480.

- $\dfrac{\text{number of minutes for Concorde flight}}{\text{number of minutes for subsonic flight}}$ is $\dfrac{210}{480}$

Since the numbers in the ratio have the same units (minutes) it is not necessary to write the units with the ratio.

Like fractions, ratios can be written in their lowest terms.

**1.** Work with a partner and write $\dfrac{210}{480}$ in its lowest terms.

**2.** How would you write the ratio if you were using hours instead of minutes as your unit of comparison?

### Vocabulary

**equivalent ratios:** Ratios that represent the same fractional number or comparison; for example, 1:2, 2:4, 3:6

**lowest terms:** A fraction or a ratio in which both the numerator and the denominator have no common factor other than 1

**ratio:** A comparison of two numbers with the same unit; for example, 1:4 means "1 compared with 4" and can be written as a fraction: $\frac{1}{4}$

**terms of a ratio:** The numbers used in a ratio

### Journal

Make notes on how ratios and fractions are similar. How are they different?

## Build Your Understanding

## Explore Equivalent Ratios

You Will Need
- scrap pieces of white paper
- red pencil

**1.** Fold a rectangular piece of paper into four equal sections as shown.

**2.** Colour one section red. Write a ratio to compare

   **a)** the red section to the white sections

   **b)** the white sections to the red section

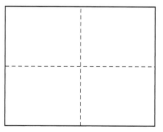

**3.** Fold the rectangle back into fours, and then fold it in half again.

**a)** How many sections are there now?

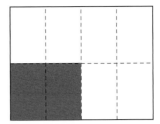

**b)** Write a ratio to compare the red with the white sections.

**c)** Write a ratio to compare the white with the red sections.

**d)** What do you notice about the ratios in question 2 and those in question 3? Are they equivalent?

**4.** By folding paper or drawing lines, show a ratio that is equivalent to 3:1 and 6:2.

## What Did You Learn?

**1.** Explain to a classmate what the following ratios mean:

**a)** 5:1      **b)** 1:5      **c)** 5:2

**2.** Write equivalent ratios for each of the following:

**a)** 2:1     **b)** 3:1     **c)** 2:5     **d)** 7:1

## Practice

**1.** Work with a partner. Flip a coin 10 times. Record each time a head turns up and each time a tail turns up. Tally the number of heads and the number of tails.

**2.** Write the following as ratios in their lowest terms:

**a)** the number of times you obtained tails compared with the number of times you obtained heads

**b)** the number of times you obtained heads compared with the number of times you obtained tails

Give a ratio in lowest terms for each of the following, based on these illustrations.

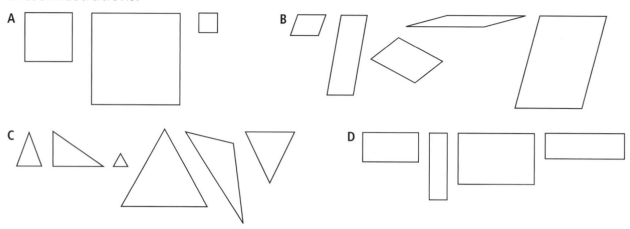

**3.** the number of squares compared with the number of parallelograms

**4.** the number of rectangles compared with the number of triangles

**5.** the number of triangles compared with the number of squares

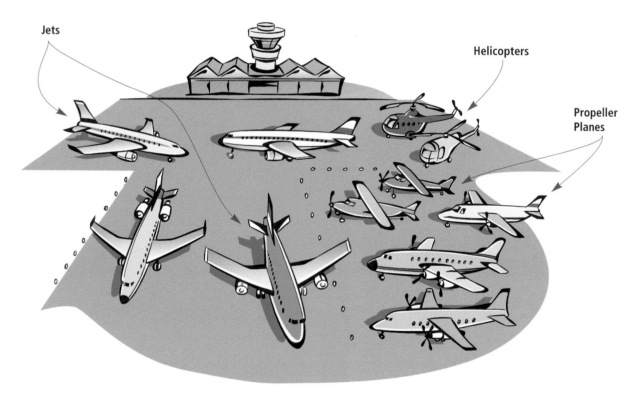

Use the airport illustration shown above to give a ratio for each flier shown. Make sure each ratio is in lowest terms.

**6.** jets compared with propeller planes

**7.** propeller planes compared with helicopters

**8.** jets compared with helicopters

**9.** Show two of the ratios you made in questions 6 to 8 as fractions.

Draw pictures in your notebook to show these ratios (do not label your pictures):

**10.** 3 : 2 **11.** 5 : 7 **12.** 12 : 17 **13.** 23 : 31 **14.** 15 : 27

**15.** Share your pictures with a classmate. Challenge your classmate to match each picture with the correct ratio.

## Technology

Using a draw application or desktop publisher, divide the work area into quarters. In the first quadrant, stamp two different small objects. Copy and paste each of these objects several times so that you have more of one object than the other. Use the text tool to make some ratio statements about the objects. For example, on the screen, Ruth stamped 4 stars and 8 hearts in the first quadrant. Under the objects Ruth could write:

1) The ratio of stars to hearts is 4:8 or $\frac{4}{8}$. This means that for every 4 stars, there are 8 hearts. The equivalent ratio is 1:2

2) The ratio of hearts to stars is 8:4 or $\frac{8}{4}$. This means that for every 8 hearts, there are 4 stars. The equivalent ratio is 2:1

Complete the other 3 quadrants in a similar way.

**Lesson 9**

# Exploring Probability

FLIGHT PLAN

PLAN:
You will learn about probabilities and possibilities.

DESCRIPTION:
When you are planning a trip, weather forecasts containing probability information can affect decisions from what to wear or pack to whether your plane takes off.

## Get Started

Probability is the measurement of the chance that an event will occur. The probability of an event is usually written as a proper fraction, but it can also be written as a percent, decimal, or ratio. To find the proper fraction for expressing a probability, use the following equation:

$$\text{Probability} = \frac{\text{number of successful outcomes}}{\text{number of all possible outcomes}}$$

For example, when you flip a coin, heads or tails are the possible outcomes. If you want the outcome to be tails, that would be the favourable outcome.

$$\text{Probability} = \frac{(1)\ \text{tails}}{(2)\ \text{heads or tails}}$$

We say the probability of tails is $\frac{1}{2}$, "1 in 2," or 50%. The probability of flipping the coin and getting heads is the same.

If you flip two pennies, the number of possibilities increases. Suppose you wanted to find the probability of flipping two pennies and obtaining two tails. First, you must identify all possible outcomes.

In this coin-flipping situation, the probability is as follows:

$$\text{Probability} = \frac{(1)\ \text{two tails}}{(4)\ \text{two heads; or two tails; or one head, one tail; or one tail, one head}}$$

We say the probability of two tails is $\frac{1}{4}$, "1 in 4," or 25%.

Work with a partner to find
• the probability of obtaining a head and a tail
• the probability of obtaining two heads
• the probability of obtaining two tails

The greater the numerator, the more likely an event will occur. When something definitely will not happen, we say the event has a 0 or 0% probability of happening. When something likely will happen, we might say it has a probability greater than $\frac{1}{2}$ or a 50% probability of happening. And when something definitely will happen, we say it has a $\frac{1}{1}$ or 100% probability of occurring.

**Vocabulary**

**frequency table:** A table that shows the number of times an event occurs
**outcome:** A possible result of a probability experiment
**possibility:** All the outcomes that could occur; for example, when flipping a coin, two outcomes are possible: heads or tails; when rolling a number cube, six outcomes are possible: 1, 2, 3, 4, 5, or 6
**probability:** The chance or the likelihood that an event will happen
**tree diagram:** A branching diagram that shows all possible outcomes of an event

**Booklink**

**Do You Wanna Bet? Your Chance to Find Out About Probability** by Jean Cushman and Martha Weston (Clarion Books: New York, NY, 1991). Two characters in this story become involved in some probability situations. For example, two invitations for birthday parties are for the same day. What is the chance that the two friends would have the same birthday? Read this book to find out!

### Experiment With Probability

You Will Need
• penny

Eric and Li flipped a coin 10 times and recorded their results in a table.

| Event | Tally | Frequency |
|---|---|---|
| heads | ⊞⊞ I | 6 |
| tails | IIII | 4 |

They used this equation to find the probability of obtaining tails:

$$P\text{ (tails)} = \frac{4 \text{ (number of tails)}}{10 \text{ (number of trials)}}$$

Eric and Li found the probability of obtaining tails to be $\frac{4}{10}$ or $\frac{2}{5}$.

### Flip a Coin

**1.** Work with a partner. Create a table like Eric and Li's.

**2.** Flip a penny 10 times and tally the times it lands on heads and the times it lands on tails. Calculate the frequency of each.

**3.** Calculate the probability of a penny landing on tails in lowest terms.

**4.** Discuss your results with another pair. Add their results to yours.

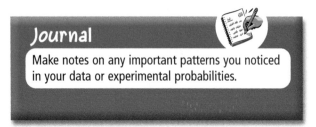

**Journal**

Make notes on any important patterns you noticed in your data or experimental probabilities.

**Technology**

Visit some Web sites that deal with probability. For example:
http://math.rice.edu/~lanius/domath/dice.html
http://nces.ed.gov/nceskids/probability/
http://www.mathgoodies.com/lessons/vol6/certain_impossible.html
http://www.harcourtschool.com/activity/probability_circus
Try some of the probability challenges. What other activities could you and a partner create that deal with probability?

## What Did You Learn?

1. How did the probability you and your partner found compare with the probability after you added another pair's results?

2. Can probability be less than or greater than 1? Explain.

3. In a few sentences, explain the differences between "possibility" and "probability." How are possibility and probability related?

4. If you flipped a penny 100 times, how many times would you expect it to come up tails? Test your prediction.

## Practice

You Will Need
• number cubes

In everyday life, we might use these words or phrases to describe the probability of events taking place.

- almost certain
- possible
- always
- likely
- never
- unlikely
- probable
- improbable

1. Write examples of events in your own life that could be described by the probability words or phrases given above. For example
   • If I play my position well in field hockey, it is possible that I will score a goal.
   • I am afraid of heights, so I will never be an astronaut.

2. Imagine that there is a 75% chance that a snowstorm will happen. Would a flight be cancelled? Explain why or why not.

3. If there is a $\frac{1}{4}$ chance of rain, would you pack an umbrella for your vacation? Why or why not?

4. What is the probability of rolling a 5 on any toss of a number cube? Explain your reasoning.

5. Is the likelihood of rolling a 5 on a toss of a number cube greater than or less than the probability of flipping a penny and getting heads? Explain why or why not.

# Lesson 10

# Managing Data

FLIGHT PLAN

PLAN:
You will learn how to create a good survey. Then you will conduct a survey and collect, organize, display, and analyze the data.

DESCRIPTION:
People depart from airports at different times of the year bound for a variety of Canadian destinations. The people who create airline schedules must know the travel preferences of Canadians. They conduct surveys to collect information, then organize and analyze the collected data.

## Get Started

You can take a survey in which you ask people in what season they would most like to travel. The entire group you want to find information about is called the population. Here the population is all Grade 6 students in Canada.

Since it is not possible to ask every Grade 6 student when he or she would most like to take a vacation, you need to survey a part of the population, or a sample, such as Grade 6 students in your class. Your sample must have the characteristics you would expect to find in the entire population.

1. With a partner, identify a good sample for your survey. Explain why the sample you selected is a good one.

   Think of a clear survey question.

**Tip**
Certain types of questions, such as yes/no or multiple choice questions, are good for surveys.

When creating questions, it is a good idea to first try them on other people. Then, if necessary, you can revise the questions to make them as clear as possible.

**2.** Predict what the results to the survey question will be.

| If You Had a Week's Vacation... When Would You Take It? | | |
|---|---|---|
| | Tally | Frequency |
| Winter | ||||| || | 7 |
| Spring | |||| | 4 |
| Summer | ||||| ||||| || | 12 |
| Fall | || | 2 |

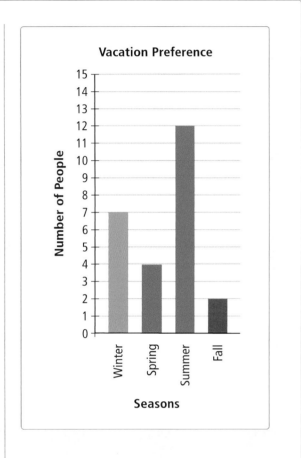

Vacation Preference

Once you have collected all your data, the next important stage is organizing the results so that you can interpret and analyze your data. A useful organizer is a frequency table, such as the one shown. In this frequency chart, seven Grade 6 students preferred to take their week-long vacation in winter.

When comparing data, it is helpful to display results using charts or graphs. The type of graph or chart you select will depend on your data. Here, students decided to use a bar graph to display the comparison results.

**3.** Discuss the following questions with a partner.
   **a)** What is the most popular season of the year among Grade 6 students to take a vacation?
   **b)** What is the least popular season?

**4.** What is the range in the data set for the frequency? (Use the Vocabulary box on the next page to help you.)

**5.** Once you have collected and organized your data, gather in groups of four to discuss these questions:

   **a)** How did your results compare with your predictions?

   **b)** What data surprised you? Why?

   **c)** What patterns do you notice?

   **d)** What conclusions can you draw from these patterns?

   **e)** What decision or decisions could you make based on the data you collected, interpreted, and analyzed? Explain.

## Vocabulary

**bias:** An emphasis on characteristics not typical of a whole population; for example, if you were trying to find the most popular television program among Grade 6 students and only surveyed boys, your survey would be biased because it ignored girls' preferences

**data:** Facts or information such as statistics

**population:** The whole group of objects or individuals considered for a survey, such as all Grade 6 students in Canada

**range:** The difference between the greatest and least values in a group; for example, in the data set 7, 5, 8, 12, the range is 12–5 = 7

**sample:** A part of a population selected to represent the population as a whole

**survey:** A method of gathering information by asking questions and recording people's answers

**tally chart:** An organizer that uses tally marks to count data and record frequencies

---

## Build Your Understanding

## Design a Survey and Analyze Data

**You Will Need**
• grid paper

Work in a small group. Your job is to collect, organize, and analyze data about holiday destination preferences for Grade 6 students. In your group, discuss and make notes to answer the following questions.

## Collect Data

**1.** Identify your survey sample.

   **a)** Who will your population be?

   **b)** Who will your sample be? How well does this sample represent the population?

   **c)** How will you avoid bias? How might you make your sample more representative of the population?

**2.** Create a good travel survey question that will provide useful and measurable results. What type of question will you use—multiple-choice, true or false, fill-in-the-blank, or some other type? Why?

**Tip**

Give the people who will complete your survey a reasonable number of destination alternatives, such as 5 or 10. Otherwise, each respondent (the person who answers the question) might have a different answer and it would be impossible to make helpful comparisons.

**Journal**

Make note of useful survey questions. You might find sample surveys at restaurants or on airplanes, or you may receive them in the mail. Explain briefly why some questions are useful survey questions and why others are not.

**3.** Try your question on a few classmates to check whether it is clear and will provide useful data. How can you improve your survey question?

**4.** Predict what you think the results of your survey will be and record your prediction.

## Organize Data

**5.** Design a frequency table with descriptive titles and column headings. Tally the responses and then determine their frequency.

**Technology**

Consider using a word processor to draft the survey question. The technology will be useful, especially if you need to revise your question several times. Some spreadsheet programs can be used to create frequency tables.

**6.** Display your results.
- What is the best graph to use when comparing your data—a bar graph? a circle graph? a line graph? Why?

## Analyze Data

**7.** Carefully examine your data, frequency table, and graph and discuss these questions with your group members:

    **a)** What is the range in your data? How did you find the range?

    **b)** How did your results compare with your predictions?

    **c)** What data surprised you? Why?

    **d)** What patterns do you notice in your data?

    **e)** What does your comparative graph show?

**8.** As a group, write a report, about two paragraphs long, in which you summarize the main conclusion or conclusions of your survey. Explain how the information can be used. What decisions could be made based on the data? Why?

## What Did You Learn?

**1.** What is the population you want to learn about in your survey?

**2.** Who is your survey sample?

**3.** Are there any possible biases in your survey sample? If so, what are they?

**4.** What makes you confident that any trends (patterns) you identified in your survey results show what the whole group, or population, prefers?

## Practice

**1.** Design a survey about a topic that you find interesting.

    **a)** Identify your survey sample.

    **b)** Create a survey question.

    **c)** Design a frequency table to organize the results.

**2.** Survey your respondents and record the results in your frequency table.

**3.** Graph the data using a graph of your choice. Why did you choose the graph that you did?

**4.** Write three questions that a classmate can answer by reading your graph.

## Show What You Know

**1. a)** Which letters in the alphabet are used most often? Which letters are used least often? Look in your local newspaper and read two or three paragraphs. Make a tally chart like this one to keep track of the letters you see.

| A | B | C | D | E | F | G | H | I | J | K | L | M |
|---|---|---|---|---|---|---|---|---|---|---|---|---|
| //// | // | | | | | | | | | | | |

| N | O | P | Q | R | S | T | U | V | W | X | Y | Z |
|---|---|---|---|---|---|---|---|---|---|---|---|---|
| | | | | | | | | | | | | |

**b)** What other data might you keep track of in this way?

**2.** Roll two number cubes 20 times. Add the total of the two number cubes each time. Which number appears most often? Which number appears least often? Create a chart to organize your information.

Chapter

# Chapter Review

**1.** Copy and complete this chart.

| Relationship to Metre | Prefix | Linear Measure | Symbol |
|---|---|---|---|
| | kilo | kilometre | km |
| 1 | - | metre | |
| | | decimetre | |
| | | centimetre | |
| | | millimetre | |

**2.** Convert the following measurements:

**a)** 3 m = ▇ cm    **b)** 500 cm = ▇ mm

**c)** 45 dm = ▇ m    **d)** 5000 km = ▇ m

**3.** Calculate the perimeter of the shapes shown below.
Show your work.

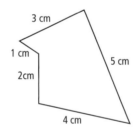

**4.** Calculate the area of the shapes shown below.
Assume each square on the grid is 1 cm². Show your work.

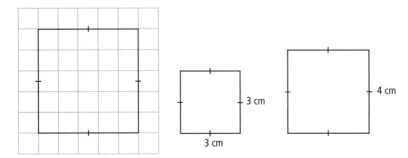

**5.** Draw a rectangle with a perimeter of 24 cm and an area of 32 cm².

**6.** Calculate the area of the parallelogram and then for each triangle shown below. What are the formulas that you used?

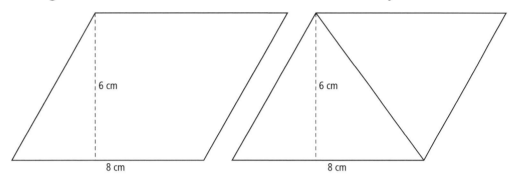

6 cm

8 cm

6 cm

8 cm

**7.** List all the pairs of factors of these numbers:

**a)** 20          **b)** 11          **c)** 48

**8.** Look at the illustration shown below. Make sure the ratios are in their lowest terms and compare the following:

**a)** the number of red squares with the number of yellow squares

**b)** the number of yellow squares with the number of red squares

**c)** the number of orange squares with the number of green squares

**d)** the number of green squares with the number of red squares

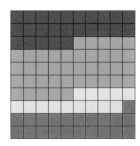

**9.** Write each ratio from question 8 as a fraction.

**10.** You Will Need
  • penny

  **a)** Flip a penny 10 times and tally the times you obtained heads and the times you obtained tails. Determine the frequency of each event.

  **b)** Calculate the probability of obtaining a head when flipping the penny.

**11.** Record all the steps you need to follow to design a survey. Remember, you need to collect, organize, display, and analyze the information. How would you pick a survey question? What choices do you need to make? How can you analyze and display your data?

# Chapter Wrap-Up

## Create Two-Dimensional Plans for a Local Airport

You have reached the end of Chapter 5. Throughout the chapter you have learned about the following mathematics topics: metric measurement units and relationships among these units; how to calculate the perimeter and area of various polygons; ratio; probability; and how to collect, organize, and analyze data from a survey.

Your task now is to apply what you know about metric measurement and data management to create two-dimensional plans for a new airport in your community.

## Create and Conduct a Survey

1. Work in a group and design a brief survey to measure where residents in your community would like to place a new airport. Identify the survey population and a reliable, unbiased sample.

2. When drafting the survey question, give people about five choices. In listing location options, consider needed and available land for an airport, convenient distance from where people live, environmental and noise concerns, and anything else you think is important.

3. Conduct the survey, organize your results, and display these results in the most appropriate graph for the type of data you collected.

4. Analyze your data and find the most popular airport location among residents.

## Identify Needed Materials

**5.** Identify the tools you will need for measuring, designing, and calculating when you create your airport plan. For example, you might need a protractor when making angle drawings.

**6.** Identify technology that could help you. For example, graphic or design software could be useful. List programs that would be helpful, that are available, and that you know how to use or could learn how to use for this project.

## Create Plans

**7.** Sketch the shape of the entire proposed airport site. Label its dimensions, perimeter, and area in the most appropriate metric unit. Draw your picture to scale and give the ratio of the picture to its actual size. You may need to do research to determine what the actual size of the building might be.

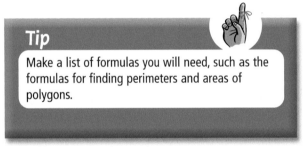

**Tip**

Make a list of formulas you will need, such as the formulas for finding perimeters and areas of polygons.

**8.** Position the runways around the airport. Give the dimensions, perimeter, and area of each runway. Make runways appropriate sizes for the aircraft that will use the airport. You may need to do research to determine these sizes.

**9.** Draw two-dimensional pictures of other important airport buildings on the site, such as the terminal and control tower. Consider shapes you have learned about in this chapter, such as rectangles, squares, triangles, and parallelograms, when creating building designs.

**10.** You might use the library or Internet to research Canadian and international airports for design ideas.

**11.** Share your plans with the class in a brief oral presentation.

# Math and Flight: Past and Present

Aviation has a long history in Canada. Math is an important part of flight.

In this chapter, you will

- estimate and calculate as you solve problems using percents
- develop your addition, subtraction, multiplication, division, and estimation skills
- identify and create nets for prisms, pyramids, and cubes
- classify pyramids
- estimate and find volume and capacity

- estimate and measure mass
- read and plot coordinates

You will then use what you have learned about polyhedrons to design and test different wing shapes.

Answer these questions to prepare for your adventure:

1. Explain in your own words what a net is. How might you create a net, and what can it be used for?

2. Explain the difference between volume and capacity.

## Lesson 1

# Estimating and Calculating Percent

FLIGHT PLAN

PLAN:
You will estimate and calculate percents, then solve problems involving percents.

DESCRIPTION:
Airlines must keep track of the number of seats that are occupied or empty for each flight. To determine if more seats can be sold, airlines might look at the percent of seats that are occupied or the percent of seats that are empty.

## Get Started

How many balls are there in this illustration? How many of the balls are shaded yellow? Just by looking, you can tell that $\frac{1}{2}$ of the balls are shaded yellow.

Percent is used to show a number out of 100. To change the fraction $\frac{1}{2}$ into a percent, you make the denominator 100. Whatever you do to the denominator, you must also do to the numerator.

$$\frac{1}{2} \times \frac{50}{50} = \frac{50}{100} = 50\%, \text{ or "50 out of 100"}$$

1. Check this result by counting the balls in the illustration.

   You can also write 50% as a decimal.

   $\frac{50}{100} = 0.50$

   50% can also be written as a ratio, 50:100.

(illustration showing numbered balls 1–100 arranged in a grid, with alternating balls shaded)

### Booklink

**Percents and Ratios** by Lucille Caron and Philip M. St. Jacques (Enslow: Berkeley Heights, NJ, 2000). This book explains the meanings of percents and ratios, and many examples help you practise.

**2.** On a 10 x 10 square grid, colour $\frac{3}{4}$ of the squares red.

$\frac{3}{4} \times \frac{25}{25} = \frac{75}{100}$

$= 75\%$

or with a calculator: $3 \div 4$

$3 \div 4 = 0.75$

$0.75 \times 100 = 75\%$

**3.** On a 10 x 10 grid, shade $\frac{1}{4}$ of the squares.

**a)** Write $\frac{1}{4}$ as a percent.

**b)** Write the percent as a decimal.

**c)** Write the percent as a ratio.

**4.** Estimate how much of the grid is shaded. Express your estimate as a percent, a fraction, and a decimal.

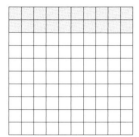

When estimating percent, you can

- compare the part to the whole or 100%
- compare the part to fractions such as $\frac{1}{4}$, $\frac{1}{2}$, or $\frac{3}{4}$
- compare the part to a part for which you know the percent

## Journal

Which percent estimation strategy do you find most useful? Why?

---

## Build Your Understanding

### Explore Percent

You Will Need
- grid paper
- set of coloured pencils

Work with a partner.

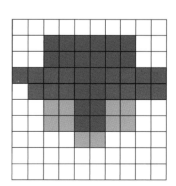

**1.** Estimate the percent of the grid that is red, green, blue, and unshaded.

**2.** Find the actual percents by counting and calculating.

**3.** Express each percent as a fraction, a decimal, and a ratio.

**4.** Shade a grid that is 35% red, 25% green, and 5% blue, and leave the rest of the squares unshaded. What percent of the squares are unshaded?

**5.** Express each percent as a fraction, a decimal, and a ratio.

**6.** Shade a grid in different colours to create a picture. List the percent of each colour. Write each percent as a fraction, a decimal, and a ratio.

**7.** Exchange your design with another pair of classmates. Ask them to estimate the percent of each colour in your design. Then, have them count squares and calculate the percents. Ask them to write each percent as a fraction, a decimal, and a ratio.

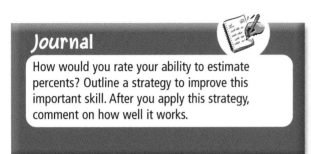

## Journal

How would you rate your ability to estimate percents? Outline a strategy to improve this important skill. After you apply this strategy, comment on how well it works.

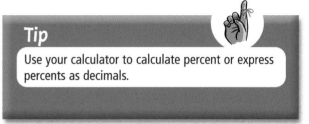

## Tip

Use your calculator to calculate percent or express percents as decimals.

## What Did You Learn?

**1.** With your partner write a definition of percent. Use an example and draw a picture to illustrate your explanation. Share and discuss your definition with another pair of students.

**2.** How are percents and fractions similar? How are they different?

**3.** How are percents and decimals similar? How are they different?

**4.** Draw a circle. Shade $\frac{1}{5}$ of it blue. What percent of the circle did you shade? What percent of the circle is unshaded?

Convert the following ratios to percents:

**1.** 32:100    **2.** 2:5    **3.** 3:4    **4.** 25:100

Convert the following percents to fractions in their lowest terms:

**5.** 33%    **6.** 25%    **7.** 50%    **8.** 75%

Convert the following fractions to percents:

**9.** $\frac{2}{5}$    **10.** $\frac{12}{25}$    **11.** $\frac{3}{4}$    **12.** $\frac{1}{2}$    **13.** $\frac{95}{100}$

Convert the following ratios to fractions in their lowest terms:

**14.** 75:100    **15.** 50:100    **16.** 8:10    **17.** 2:4

**18.** Copy and complete this conversion chart.

| Percent | Fraction | Decimal |
|---|---|---|
| 100% | | |
| | | 0.45 |
| | | 0.86 |

## Math Problems to Solve

**19.** Imagine that an airplane has 100 seats. There are 25 rows with 4 seats in each row. The shaded squares represent the seats that are occupied by passengers.

   **a)** Estimate the percent of the seats that are occupied.

   **b)** Count squares to calculate the percent of seats that are occupied.

   **c)** What percent of seats are empty?

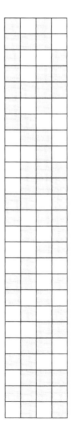

**20.** Imagine that an airplane can hold 50 passengers. On the last five flights, the numbers of passengers were as follows:

**a)** Flight 1: $\dfrac{35}{50}$

**b)** Flight 2: $\dfrac{45}{50}$

**c)** Flight 3: $\dfrac{27}{50}$

**d)** Flight 4: $\dfrac{39}{50}$

**e)** Flight 5: $\dfrac{21}{50}$

At least 70% of the seats must be occupied for the airline to make a profit. Determine which flights were profitable. Show your work in numbers, pictures, and words.

**Journal**

Describe in numbered steps the strategy you will apply to solve the problem. What will you do first, second, and so on? In what other situations could you apply this problem-solving strategy?

## Lesson 2

# Operations

### FLIGHT PLAN

PLAN:
You will improve your adding, subtracting, multiplying, dividing, and estimating abilities.

DESCRIPTION:
To prepare for a flight, the pilot of a small plane might add the mass of passengers, fuel, and cargo (supplies) and subtract items that are too heavy, so the plane can lift off the ground. He or she might multiply the distance the plane can travel in one hour by the number of travel hours. The pilot might divide the travel distance by travel time to determine the length of each leg of a journey.

## Get Started

### Addition

To find the sum of 687, 453, 841, and 396, arrange the numbers in columns and add them from top to bottom and right to left.

$$
\begin{array}{r}
\scriptstyle 2\,1 \\
687 \\
453 \\
841 \\
+\ 396 \\
\hline
2377
\end{array}
$$

| Add ones | Add tens | Add hundreds |
|---|---|---|
| 7 + 3 + 1 + 6 | 8 + 5 + 4 + 9 (**+ 1** carried over) | 6 + 4 + 8 + 3 (**+ 2** carried over) |

### Journal

Preview the questions in this lesson. How would you rate your ability to do these types of addition, subtraction, multiplication, and division questions? What do you do well when performing these operations? What do you need to improve on?

## Subtraction

To find the difference between 82 715 and 9867, arrange the numbers in columns with the greater number on the top. Any needed borrowing is shown in red.

```
  7 11 16 10
  82 715
 −  9 867
 ─────────
  72 848
```

| Subtract ones | Subtract tens | Subtract hundreds | Subtract thousands | Subtract ten thousands |
|---|---|---|---|---|
| 5 – 7 | 0 – 6 | 6 – 8 | 1 – 9 | 7 – 0 |
| **(borrow 1)** | **(borrow 1)** | **(borrow 1)** | **(borrow 1)** | |
| 15 – 7 | 10 – 6 | 16 – 8 | 11 – 9 | |

Check your answer by adding the difference to the number above it. Your answer should be the top number.

## Multiplication

To find the product of 517 and 47, write the greater number above the smaller number.

```
Multiply   517              Multiply   517           Add partial   20 680
           x 7                         x 40           products     + 3 619
      ─────────                   ─────────                        ───────
            49  (7 x 7)               280  (40 x 7)                 24 299
            70  (7 x 10)              400  (40 x 10)
        + 3500  (7 x 500)        + 20 000  (40 x 500)
      ─────────                   ─────────
          3619                      20 680
```

How could you check your work?

**Vocabulary**

**difference:** The answer to a subtraction question
**digit:** Numerals from 0 to 9
**dividend:** The number being divided
**divisor:** The number that divides the dividend
**operation:** A process in mathematics performed according to specific rules, such as addition, subtraction, multiplication, or division
**product:** The answer to a multiplication question
**quotient:** The answer to a division question
**sum:** The answer to an addition question

# Division

To find the quotient when 1095 is divided by 28, write a long division sign and place the dividend underneath and the divisor to the left. Draw vertical lines to show the place-value columns in the dividend:

Mentally round the divisor to the nearest 10 (30, in this case). Looking left to right, ask the following questions:

**a)** How many times can 30 divide into 1? The answer is 0, so include the next place value.

**b)** How many times can 30 divide into 10? The answer is 0, so include the next place value.

**c)** How many times can 30 divide into 109? The answer is 3. Place the 3 above the last digit included (9, in our example). Multiply 28 by 3, and record the result below 109. Subtract the answer from 109.

$$
\begin{array}{r}
3\phantom{0} \\
28\overline{)1\,0\,9\,5} \\
-\,8\,4\phantom{0} \\
\hline
2\,5\phantom{0}
\end{array}
$$

**d)** Repeat the process. How many times can 30 go into 25? The answer is 0, so include the next place value.

**e)** Move down the next place value. How many times can 30 go into 255? The answer is 9. Place the 9 above the last digit included (5, in our example). Multiply 28 by 9 and record it below 255. Subtract the answer from 255.

$$
\begin{array}{r}
3\,9 \\
28\overline{)1\,0\,9\,5} \\
-\,8\,4\,\downarrow \\
\hline
2\,5\,5 \\
-\,2\,5\,2 \\
\hline
3
\end{array}
$$

**f)** How many times can 30 go into 3? The answer is 0. We do not have any more place values to include. 3 is the remainder.

Therefore, 28 goes 39 times into 1095 with a remainder of 3.

Check your answer by multiplying the quotient by the divisor and adding the remainder.

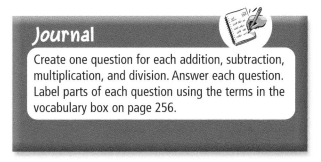

## Journal

Create one question for each addition, subtraction, multiplication, and division. Answer each question. Label parts of each question using the terms in the vocabulary box on page 256.

## Add, Subtract, Multiply, and Divide

Estimating will help you check the reasonableness of your answers. For example, to check the product in the Get Started multiplication question, you might do the following:

| | |
|---|---|
| 517 | Think: Rounded to nearest hundred, 517 is about 500. |
| x 47 | Think: Rounded to the nearest ten, 47 is about 50. |
| 24 299 | Think: I can calculate 500 x 50 = 25 000 mentally. |
| | The answer is reasonable. |

How could you apply this estimation strategy to check your sums, differences, products, and quotients? Use the strategy to check the reasonableness of your answers in the questions below.

**1.** Add the following:

| **a)** 71 | **b)** 781 | **c)** 541 | **d)** 953 | **e)** 518 | **f)** 711 |
|---|---|---|---|---|---|
| 238 | 634 | 362 | 871 | 999 | 887 |
| 576 | 918 | 432 | 265 | 396 | 492 |
| + 93 | + 356 | + 860 | + 187 | + 214 | + 566 |

**2.** Subtract the following:

| **a)** 58 716 | **b)** 62 123 | **c)** 93 856 | **d)** 84 531 | **e)** 54 333 | **f)** 48 231 |
|---|---|---|---|---|---|
| − 4 388 | − 5 796 | − 7 839 | − 6 896 | − 1 997 | − 6 941 |

**3.** Multiply the following:

| **a)** 481 | **b)** 192 | **c)** 496 | **d)** 899 | **e)** 566 | **f)** 722 |
|---|---|---|---|---|---|
| x 37 | x 81 | x 78 | x 37 | x 94 | x 53 |

**4.** Divide the following:

**a)** 9842 ÷ 27    **b)** 58$\overline{)3455}$    **c)** 5130 ÷ 32

**d)** 75$\overline{)1262}$    **e)** 24$\overline{)7864}$    **f)** 8756 ÷ 86

**Tip**

Use a calculator to check your answers.

**Journal**

Review the journal entry in which you listed your computation strengths and areas needing improvement. How have your skills changed?

## What Did You Learn?

**1.** What relationships do you notice among operations? For example, how do you use multiplication and subtraction when dividing numbers?

**2.** Meet with a partner. Refer to the operations questions you completed to explain the following math terms in your own words: difference, sum, quotient, product, estimation, borrowing, and carrying over.

## Practice

Estimate the best answer to each question below.

|  | A | B | C |
|---|---|---|---|
| 1. 259 + 150 | 500 | 400 | 450 |
| 2. 396 − 281 | 100 | 150 | 200 |
| 3. 587 x 32 | 20 000 | 18 000 | 19 000 |
| 4. 411 ÷ 38 | 12 | 11 | 10 |

Add the following:

| **5.** | **6.** | **7.** | **8.** | **9.** | **10.** |
|---|---|---|---|---|---|
| 971 | 432 | 921 | 995 | 668 | 611 |
| 436 | 834 | 667 | 861 | 899 | 387 |
| 576 | 488 | 591 | 365 | 696 | 491 |
| + 183 | + 396 | + 255 | + 486 | + 715 | + 831 |

Subtract the following:

| **11.** | **12.** | **13.** | **14.** | **15.** | **16.** |
|---|---|---|---|---|---|
| 27 237 | 92 423 | 73 858 | 64 535 | 14 353 | 78 431 |
| − 5 368 | − 7 799 | − 6 837 | − 5 192 | − 9 997 | − 2 941 |

Multiply the following:

| **17.** | **18.** | **19.** | **20.** | **21.** | **22.** |
|---|---|---|---|---|---|
| 561 | 292 | 597 | 800 | 869 | 999 |
| x 75 | x 62 | x 81 | x 36 | x 27 | x 63 |

Divide the following:

**23.** 1842 ÷ 17

**24.** $38\overline{)7495}$

**25.** 6137 ÷ 52

**26.** 9262 ÷ 34

**27.** $64\overline{)8864}$

**28.** 8815 ÷ 96

## Show What You Know

### Review: Lessons 1 and 2, Percent and Operations

**1.** Claire found out that her soccer team needs to win 80% of their games to get a championship trophy at the end of the season. They have played 20 games so far, and have won 15.

   **a)** What percent of the games have they won so far? Is this enough to win the championship trophy? Why or why not?

   **b)** Claire's team has 30 more games to play this season. If they win 25 more games, will they receive the championship trophy? Explain how you solved this problem.

**2.** There are three planes, and the maximum amount each plane can carry is listed in the following chart:

| Plane | Maximum Amount |
|-------|----------------|
| A | 36 868 kg |
| B | 9 000 kg |
| C | 17 486 kg |

Calculate the difference in maximum amounts between Plane A and Plane B.

Calculate the difference in maximum amounts between Plane C and Plane B.

Which of the three planes would be the best choice to carry the following amounts?

   **a)** 132 pieces of luggage, each with an average mass of 36 kg, and 132 passengers, each with an average mass of 72 kg

   **b)** 101 pieces of luggage, each with an average mass of 28 kg, and 101 passengers, each with an average mass of 60 kg

   **c)** 280 pieces of luggage, each with an average mass of 37 kg, and 280 passengers, each with an average mass of 74 kg

## Lesson 3

# Exploring Nets and Prisms

*FLIGHT PLAN*

*PLAN:*
**You will identify and create nets for various three-dimensional prisms.**

*DESCRIPTION:*
*Three-dimensional shapes are very important to flight. Some shapes have flat surfaces that resist the flow of air. Others, such as this airfoil, use air to their advantage. The airfoil allows air to flow over the wing, which helps lift the plane into the air.*

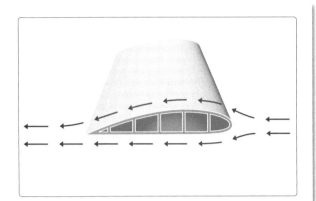

## Get Started

A polyhedron is a three-dimensional solid with polygon faces. In a polyhedron called a prism, the end faces of the solid figure are congruent and parallel.

Prisms are named for their bases.

You Will Need
• box that forms a rectangular prism
• scissors
• grid paper
• adhesive tape

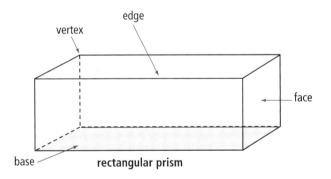

Work with a partner or in a small group.

1. Carefully take the box apart so that it lies in one flat piece.

2. Draw a sketch on grid paper of the flat pattern used to make the three-dimensional box.

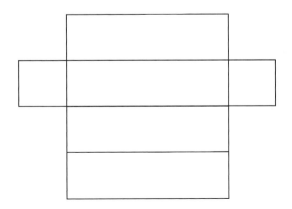

**base:** A side of a polygon or face of a solid figure by which the figure is measured or named. The symbol for base is *b*.

**congruent:** Exactly the same size and shape (the symbol for congruence is ≡ )

**edge:** The line segment where two faces of a three-dimensional figure meet

**face:** One of the polygons of a solid figure

**hexagon:** A polygon with six sides

**net:** A two-dimensional pattern of a three-dimensional figure

**octagon:** A polygon with eight sides

**parallel lines:** Two lines that are continuously the same distance apart (they never cross)

**pentagon:** A polygon with five sides

**plane of symmetry:** A plane that divides a solid into two congruent solids

**polygon:** A shape that has at least three straight sides

**polyhedron:** A three-dimensional figure with polygons for its faces

**prism:** A three-dimensional figure with two faces that are congruent and parallel and other faces that are quadrilaterals

**rectangular prism:** A three-dimensional figure in which all six faces are rectangles; a square prism is a rectangular prism with two square faces

**three-dimensional:** Having three dimensions: length, width, and height or depth

**two-dimensional:** Having two dimensions: length and width but not height

**vertex:** The point where two rays of an angle or two edges meet in a plane figure, or the point where three or more sides meet in a solid figure (the plural of vertex is vertices)

3. Cut out the flat pattern you drew on grid paper, fold it, and tape it to form a rectangular prism.

4. Are there other nets that could be used to create a rectangular prism? Explore your ideas with your partner or group.

### Tip

You created a net. A net is a flat shape that can be folded to create a three-dimensional object.

## Build Your Understanding

## Draw and Construct Prisms From Nets

You Will Need
- paper
- scissors
- glue or tape
- grid paper

Work with a partner.

1. Explore the rectangular prism you created from a net in Get Started.

2. Make a larger version of the net shown here and cut it out. Fold and tape the pattern to create a triangular prism.

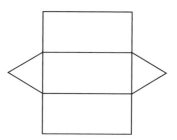

**3.** Select another prism from those shown below.
Make a larger version of the net and cut it out.
Fold the pattern and tape it to create the prism.

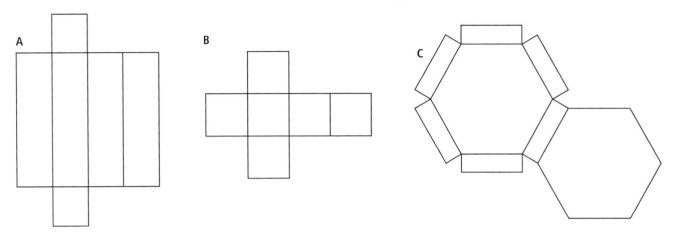

A

B

C

**4.** Copy the chart shown below into your notebook and complete
it for three more prisms.

| Name of Prism | Three-Dimensional Sketch of Solid | Number of Faces | Number of Edges | Number of Vertices | Sketches of Faces |
|---|---|---|---|---|---|
| | | | | | |
| | | | | | |
| | | | | | |
| | | | | | |

**5.** Answer these questions in your journal:

**a)** Why do you think the prisms are called triangular and
rectangular prisms?

**b)** Are faces within a prism congruent? If so, which faces are
congruent? Which faces are parallel?

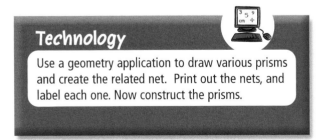

**Technology**

Use a geometry application to draw various prisms
and create the related net. Print out the nets, and
label each one. Now construct the prisms.

## What Did You Learn?

**1.** Define a "net" in your own words. Use a labelled sketch to support your definition.

**2.** What patterns did you notice in your chart results? What conclusion(s) can you draw from these patterns?

**3.** Match the prisms in the left column to the net in the right column that was used to create the prism.

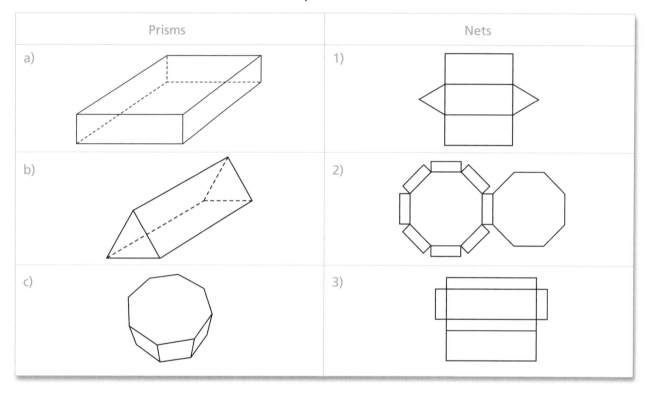

| Prisms | Nets |
|---|---|
| a) | 1) |
| b) | 2) |
| c) | 3) |

**1.** Draw nets for each of these three-dimensional shapes:

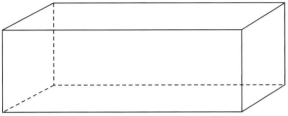 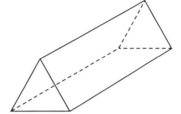 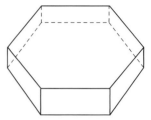

**a)** rectangular prism      **b)** triangular prism      **c)** hexagonal prism

**2.** How many faces, edges, and vertices does each shape in question 1 have? Use the nets you drew to help you.

**3.** Explain to a classmate the difference between two-dimensional and three-dimensional.

**4.** Draw nets for three prisms of your choice. List the name of each prism on a separate piece of paper. Challenge a classmate to match each net with the correct prism name.

## Extension

**5.** A plane of symmetry divides a solid into two congruent solids. Predict how many planes of symmetry each prism in the lesson has. Then, design a way to test your predictions. Share, compare, and discuss your findings in a small group.

**Tip**

You might consider using modelling clay if you want to slice prisms to find their planes of symmetry.

**Lesson 4**

# Classifying Pyramids

*FLIGHT PLAN*

*PLAN:*
You will classify pyramids according to their vertices, edges, and faces. As well, you will draw nets for various types of pyramids and use the nets to construct three-dimensional figures.

*DESCRIPTION:*
One of the earliest designs for a parachute was drawn by the great Renaissance artist, engineer, and inventor Leonardo da Vinci (1452–1519). Da Vinci's original idea for the parachute was a pyramid shape. Today parachutes come in many shapes.

## Get Started

Leonardo da Vinci's parachute design was a square-based pyramid.

1. Look closely at da Vinci's parachute design and explain to a classmate why it might be called a square-based pyramid.

2. With a partner, identify common characteristics of the two types of pyramids shown on the right. How are they different?

3. Compare these pyramids with the prisms you explored in Lesson 3. What do they have in common? How are they different?

**Tip**

A pyramid is a solid figure with a polygon base and triangular faces that meet at one common point.

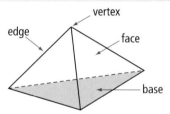

**triangular-based pyramid**

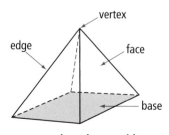

**square-based pyramid**

**4.** Predict what the nets for a triangular-based pyramid and a square-based pyramid will look like. Sketch your ideas for the nets on grid paper. Meet with a small group of classmates to discuss your predictions. You will check your predictions and use these nets to construct triangular-based and square-based pyramids.

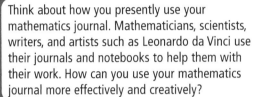

**Journal**

Think about how you presently use your mathematics journal. Mathematicians, scientists, writers, and artists such as Leonardo da Vinci use their journals and notebooks to help them with their work. How can you use your mathematics journal more effectively and creatively?

**Vocabulary**

**base:** A side of a polygon or a face of a solid figure by which the figure is measured or named. The symbol for base is *b*.
**congruent:** Exactly the same size and shape
**pyramid:** A solid figure with a polygon base and triangular faces that meet at a common vertex
**regular polyhedron:** A polyhedron in which all the faces of the figure are congruent; for example, in a triangular-based pyramid, all faces are congruent equilateral triangles

## Build Your Understanding

## Draw and Construct Pyramids From Nets

Alexander Graham Bell invented a special kind of kite that was lighter than other kites, but much stronger and more powerful. Instead of simply making a larger wing, he joined many smaller wings together in the shape of a triangular-based pyramid. Look at the picture of his kite on page 249.

You Will Need
• triangular-based pyramid net
• paper
• scissors
• glue or tape
• grid or dot paper

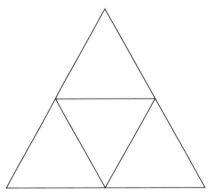

**1.** On grid or dot paper, draw the net for a square-based pyramid.

**2.** Compare the net for a square-based pyramid to this net for a triangular-based pyramid. How are they the same? How are they different?

**3.** Cut out the net for the square-based pyramid. Fold and tape the pattern to create a square-based pyramid.

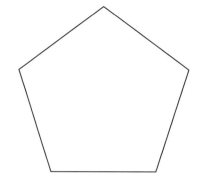

**4.** On grid paper or dot paper, draw the net for a pentagonal-based pyramid. Cut out the net. Fold and tape the pattern to create a pentagonal-based pyramid.

**5.** How many triangular faces will the pentagonal-based pyramid have? How do you know this?

**6.** Check the predictions you made in Get Started about the nets for triangular-based and square-based pyramids. How accurate were your predictions?

**7.** Copy the chart shown below into your notebook.

| Name of Pyramid | Three-Dimensional Sketch of Solid | Number of Faces | Number of Edges | Number of Vertices | Sketches of Faces |
|---|---|---|---|---|---|
|  |  |  |  |  |  |
|  |  |  |  |  |  |

**8.** Use the pyramids you constructed to complete the chart.

## What Did You Learn?

**1.** What patterns do you notice in your chart results? What conclusions can you draw from these patterns?

**2.** Compare your results in the pyramid chart with the chart you completed for different types of prisms in Lesson 3. Refer to the charts to explain important similarities in and differences between prisms and pyramids.

**3.** Predict what the chart results will be for an octagonal-based pyramid.

### Technology

Using a spreadsheet application, input pyramid data into a spreadsheet using the following headings: Type of Base and Number of Faces. Create a bar graph from the spreadsheet. What do you notice about the relationship between the type of base and the number of faces? Now input data for Number of Edges and Number of Vertices and create a bar graph for each. Use the text tool to write some sentences about your conclusions.

1. In your notebook, draw nets for the following shapes:
   a) a rectangular prism
   b) a triangular-based pyramid
   c) an octagonal-based pyramid
   d) a triangular prism

2. Draw a net for a different pyramid or prism of your choice.

3. Draw a net for a pyramid with 6 vertices.

4. Draw a net for a pyramid with 5 faces.

5. Identify the polyhedrons formed by these nets. Give reasons for your decisions.

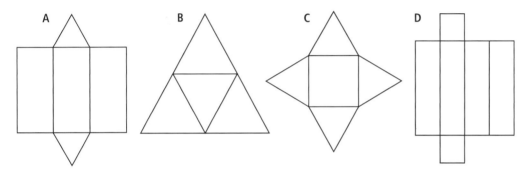

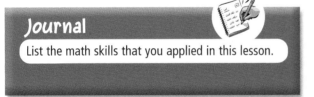

**Journal**

List the math skills that you applied in this lesson.

**Lesson 5**

# Exploring Cubes

FLIGHT PLAN

PLAN:
You will design various nets for cubes and use the nets to construct cubes.

DESCRIPTION:
People who handle cargo must move cube containers and stack them in the cargo sections of aircraft. The containers might be flown to communities within Canada or to people in different parts of the world.

## Get Started

You Will Need
• cubes

Work with a partner to explore real-world cubes, such as a number cube, a sugar cube, a child's block, or any other cube. Look closely at the cube and answer the following questions.

**1.** How many faces does the cube have?

**2.** What shape is each face?

**3.** Which faces are congruent?

**4.** Which faces are parallel?

**5.** How many edges does the cube have?

**6.** What is the angle formed where the edges meet at any vertex?

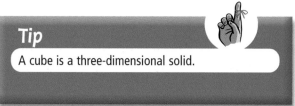

**Tip**

A cube is a three-dimensional solid.

**7.** How many vertices are there?

**8.** What kinds of symmetry does the cube have?

Like a triangular-based pyramid, a cube is a regular polyhedron—all its faces are congruent. In a triangular-based pyramid, all the faces are congruent equilateral triangles. In a cube, all the faces are congruent squares.

**9.** Sketch a net for a cube. Is there more than one possible net that can be used to create a cube? Explain.

### Journal

Make a list of cube-shaped objects. Add new cube examples as you encounter them in your mathematics learning and in your learning outside the classroom.

### Vocabulary

**cube:** A three-dimensional solid with six congruent square faces

**surface area:** The sum of areas of the faces of a three-dimensional object

## Build Your Understanding

### Predict and Test Whether Net Patterns Make Cubes

You Will Need
• scissors
• glue or tape
• grid paper

**1.** In a small group, study these patterns and predict which will be nets for cubes and which will not. Make sure you record your predictions and be prepared to explain why you think certain patterns can create cubes.

**2.** Test your predictions. Have each person in the group make a larger copy of one pattern, cut it out, then fold and tape it to create a cube. Which patterns are nets for cubes? Which are not?

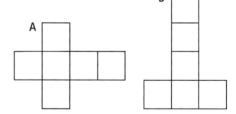

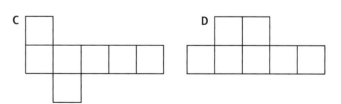

## Draw Different Perspectives of Cubes

You Will Need
- grid paper
- square dot paper
- isometric dot paper

1. Use one of the cubes you examined or constructed in this lesson as a drawing model. On grid paper, then square dot paper, and then isometric dot paper, make drawings that clearly communicate the three-dimensional qualities of the cube.

2. Which type of paper did you find most useful when drawing the three-dimensional cube? Why?

**Vocabulary**

**isometric dot paper:** Dot paper whose dot pattern is formed by the vertices of equilateral triangles; it is used to make three-dimensional drawings

## What Did You Learn?

1. Copy this chart into your notebook. Use what you learned in previous lessons and in this lesson to complete the chart.

| Name of Prism or Pyramid | Three-Dimensional Sketch of Solid | Number of Faces | Number of Edges | Number of Vertices | Sketches of Faces |
|---|---|---|---|---|---|
| triangular prism | | | | | |
| rectangular prism | | | | | |
| cube | | | | | |
| triangular-based pyramid | | | | | |
| square-based pyramid | | | | | |

2. Meet in a group to share, compare, and discuss your charts. What patterns do you notice? What conclusions can you draw from these patterns?

3. Draw and label a net for a cube that has a side length of 4 cm. Cut out, fold, and tape the net to create a cube.

1. Review the charts you made about three-dimensional figures. Create "Who Am I?" questions based on this chart information. For example, "I have six congruent square faces. Who am I?"

2. **a)** If you made a sculpture using a rectangular prism, a cube, and a square-based pyramid, how many faces would there be all together?

   **b)** How many vertices would there be all together?

   **c)** How many edges would there be all together?

   **d)** Sketch all the two-dimensional shapes that would be found in your sculpture. How many of each shape would you have?

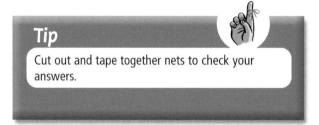

**Tip**

Cut out and tape together nets to check your answers.

3. What cube shapes can you find around home or at school? How do you know they are cubes?

## A Math Problem to Solve

4. Ahanu built a model using cubes, triangular prisms, triangular-based pyramids, and square-based pyramids. When Ahanu counted the faces, the total was 75. How many of each different three-dimensional solid might Ahanu have used?

### Show What You Know

**Review: Lessons 3 to 5, Three-Dimensional Solids**

Use small square tiles to answer these questions:

1. Begin with 3 tiles. Arrange them so that each side touches a complete side of another tile.

2. Use 4 tiles to create a shape, then 5 tiles, and then 6 tiles.

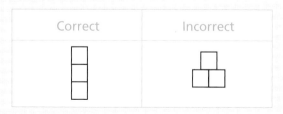

| Correct | Incorrect |
|---------|-----------|

3. Which of your tile patterns would be nets for cubes? Explain.

**Lesson 6**

# Exploring Volume

*FLIGHT PLAN*

*PLAN:*
**You will learn to estimate and find volume by counting cubes.**

*DESCRIPTION:*
How much cargo and baggage will fit into the hold of an airplane? How much fuel can an airplane's fuel tanks hold? These important flight questions can be answered by calculating volume.

## Get Started

Volume is the space an object occupies. Capacity is the amount a container will hold.

Volume is measured in cubic units. For this cube, we say the volume is "one cubic centimetre" and write it as 1 cm³.

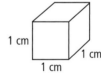

Another common metric unit for measuring volume is the cubic metre. This cube has a volume of "one cubic metre," or 1 m³.

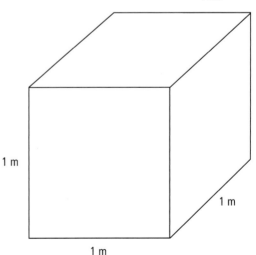

**1.** What is the volume of this rectangular prism? How do you know?

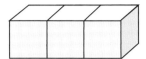

**2.** What is the length of the prism? What is its width? What is its height?

**3.** Work with a partner. What is the volume of this solid? How can you tell?

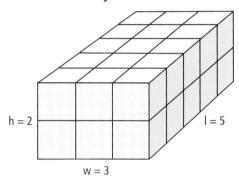

h = 2   l = 5   w = 3

**4.** Meet with a group of classmates to share, compare, and discuss strategies for finding volume.

You Will Need
- linking cubes (about 20)

**5.** In your group work, with about 20 linking cubes. Imagine that each cube has a volume of 1 cm$^3$. Use the cubes to make rectangular prisms with different dimensions. Challenge classmates to estimate and then find the volume of the solids you create.

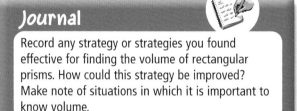

**Journal**

Record any strategy or strategies you found effective for finding the volume of rectangular prisms. How could this strategy be improved? Make note of situations in which it is important to know volume.

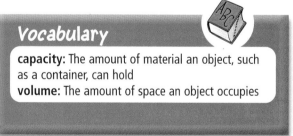

**Vocabulary**

**capacity:** The amount of material an object, such as a container, can hold
**volume:** The amount of space an object occupies

## Build Your Understanding

## Estimate and Find the Volume of Rectangular Prisms

**You Will Need**
- linking cubes (about 40)
- grid paper or isometric dot paper

**1.** Copy a chart like this one into your notebook.

 **Tip**

Include additional rows in your notebook chart so you can record the dimensions and volumes of several rectangular prisms.

| Estimated Volume | Length (l) | Width (w) | Height (h) | Actual Volume (V) |
|---|---|---|---|---|
|  |  |  |  |  |
|  |  |  |  |  |

**2.** For this rectangular prism, do the following:
   **a)** Estimate the volume and record your estimate.
   **b)** Use cubes to make the rectangular prism.
   **c)** Find the length, width, and height of the rectangular prism. Assume that each linking cube measures 1 cm x 1 cm x 1 cm, or 1 cm³.
   **d)** Use a counting strategy to find the volume.
   **e)** Record information in the appropriate chart cells.

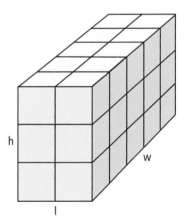

**3.** What would the volume of the prism be if you increased the height by 1 cm? Why?

**4.** Use your linking cubes to create a rectangular prism with dimensions different than any you have explored so far.
   **a)** Find the length, width, and height of the rectangular prism.
   **b)** Use a counting strategy to find the volume.
   **c)** Record relevant information in the appropriate chart cells.
   **d)** Make a three-dimensional drawing on grid paper or isometric dot paper showing a good view of your rectangular prism's three dimensions.

**5.** Meet as a large group to discuss your results. Create a class chart on which you record dimensions and volumes for all the different rectangular prisms that students created.

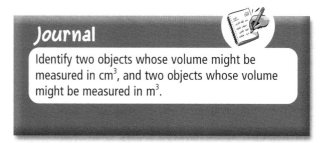

**Journal**

Identify two objects whose volume might be measured in $cm^3$, and two objects whose volume might be measured in $m^3$.

## What Did You Learn?

**1.** Look closely at the class chart. What relationship do you notice among the length, width, and height of rectangular prisms and their volumes?

**2.** Suggest a formula based on this relationship that you might use to calculate the volume of rectangular prisms.

**3.** With your cubes, create a different rectangular prism. Use your formula to calculate its volume.

**4.** Check the accuracy of your formula by counting cubes to find the volume. If necessary, revise your formula.

**1.** For each photo, identify the best measurement unit for measuring the volume of each rectangular prism.

**2.** Describe in numbered steps the strategy you would use to find the volume of each rectangular prism.

You Will Need
- linking cubes or number cubes
- grid paper or isometric dot paper

**3.** Make a rectangular prism with your cubes. Use the formula you explored in the lesson to calculate the volume. Check the accuracy of your calculations by counting cubes. Challenge classmates to find the volume of your rectangular prism.

**4.** Use grid paper or isometric dot paper to draw accurate three-dimensional pictures of rectangular prisms. Challenge classmates to calculate the volume of the solids that you drew.

**5.** Draw the net for a rectangular prism. Cut it out, then fold and tape the pattern to create a prism. Calculate the volume of the rectangular prism you created.

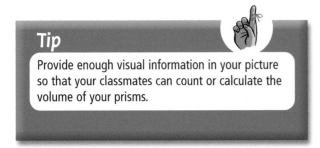

**Tip**

Provide enough visual information in your picture so that your classmates can count or calculate the volume of your prisms.

# Lesson 7

# Calculating Volume and Capacity

## FLIGHT PLAN

### PLAN:
You will estimate volume and capacity and use formulas to calculate volume and capacity. As well, you will solve volume and capacity problems.

### DESCRIPTION:
Many airport jobs involve handling luggage. Since there is limited storage room on airplanes, each passenger's bags must not take up too much space. As a result, airlines restrict the size of bags that people can take on the plane.

## Get Started

**1.** Work with a classmate. What type of polyhedron is this piece of personal carry-on luggage?

**2.** Find the area of its base.

**3.** Estimate the volume of the case and record your estimate.

**4.** Use the formula shown in the screened box to calculate the volume of the case. How close was your estimate to the actual volume?

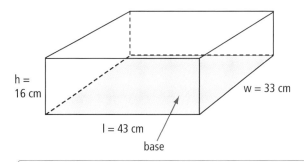

h = 16 cm

w = 33 cm

l = 43 cm

base

You can think of the relationships among a rectangular prism's dimensions, surfaces, and volume in a number of ways:

Volume = length x width x height

Volume = (area of base) x height

> The case is an example of a rectangular prism. It is three-dimensional, having the dimensions of length (*l*), width (*w*), and height (*h*). The volume (*V*) of a rectangular prism is length times width times height.
>
> Volume = length x width x height
>
> $V = l \times w \times h$

$V = L \times w \times h$

## Build Your Understanding

### Estimate and Calculate Volume

1. Estimate the volume of this carry-on bag. Use the formula for the volume of a rectangular prism to calculate its volume.

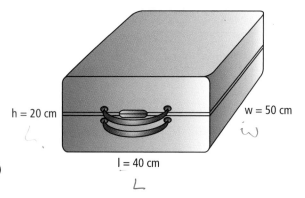

h = 20 cm
w = 50 cm
l = 40 cm

2. On an airline, passengers can check two pieces of luggage. The overall measurement of each piece (length *plus* width *plus* height) must be less than 158 cm.

   **a)** Design a suitcase that fits within the airline's overall measurement requirements. Draw a sketch of this three-dimensional suitcase on grid or isometric dot paper and label its dimensions. Use the correct formula to calculate the volume of the suitcase.

   **b)** Design a suitcase that fits within the airline's overall measurement requirements but that will have a greater volume than the suitcase you designed in part a). Draw a sketch of this three-dimensional suitcase on grid or isometric dot paper and label its dimensions. Use the correct formula to calculate the volume of the suitcase.

3. Exchange your suitcase sketches with a classmate. Have him or her estimate the volume and then use the correct formula to check your calculations.

4. What is the volume of this laptop computer?

4.5 cm
27 cm
32 cm

Volume and capacity are closely related. Volume is the amount of space an object occupies, and capacity is the amount a container holds.

5. What is the capacity of this computer carrying case?

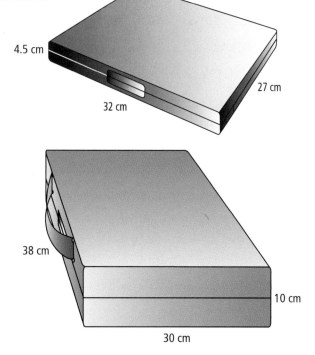

38 cm
10 cm
30 cm

**6.** What is the capacity of the aquarium? Express your answer in the following units:

**a)** cubic centimetres (cm³)

**b)** millilitres (mL)

**c)** litres (L)

40 cm

45 cm

50 cm

## What Did You Learn?

**1.** In your own words, explain the difference between volume and capacity.

**2.** Write a formula for finding the area of the base of a rectangular prism. What is the relationship between the area of the base of a rectangular prism and its volume?

**3. a)** Estimate the volume of each container shown in the illustration.

**b)** What measurements would you need to calculate the volume of these containers?

**c)** What measurement unit(s) would you use? Why?

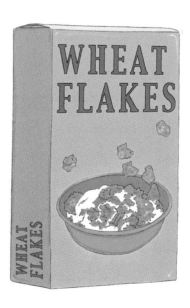

WHEAT FLAKES

ORANGE JUICE

## Practice

Calculate the volume of the following shapes:

**1.** a rectangular prism (length = 15 cm, width = 5 cm, height = 7 cm)

**2.** a cube (length = 5 m)

**3.** a rectangular prism (length = 302 mm, width = 146 mm, height = 221 mm)

**4.** In your notebook, draw a net for the shape in question 1 and label the measurements.

## Math Problems to Solve

What is the volume of the water in the aquarium? Express your answer in the following metric units:

**5.** cubic centimetres (cm$^3$)

**6.** millilitres (mL)

**7.** litres (L)

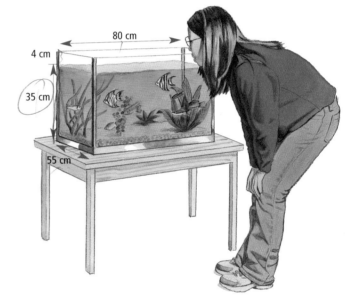

What is the capacity of the aquarium? Express your answer in the following metric units:

**8.** cubic centimetres (cm$^3$)

**9.** millilitres (mL)

**10.** litres (L)

**11.** An airliner has a wingspan of 60 m. Its length is 64 m and the height of the aircraft from the ground to tail tip is 17 m. You need to add an additional 20 m each to the length, width, and height so that the aircraft has room to move around. Give the dimensions of a hangar that could house this aircraft. What is the capacity of this building? Explain the strategy you used to solve this problem.

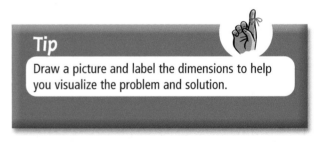

**Tip**

Draw a picture and label the dimensions to help you visualize the problem and solution.

**12.** Find a number of different-sized prisms, such as food containers or suitcases, that you have at home. Estimate the volume of each prism. Then, use the correct formula to calculate the volume of each prism.

# Lesson 8

# Measuring Mass

## FLIGHT PLAN

*PLAN:*

You will learn how to estimate and measure mass. You will also examine the relationships among metric units.

*DESCRIPTION:*

Balloon flight became possible when people learned how to use gases with lighter masses than the air around us, such as hot air, hydrogen, and helium. In 1783, two Frenchmen, Joseph and Étienne Montgolfier, made the first passenger flight in this hot air balloon. They travelled only a few kilometres before landing.

## Get Started

Mass is the amount of matter in an object. Mass is different from weight, which is a measure of the force of gravity acting on an object. Mass never changes; it is the same no matter where the object is in the universe.

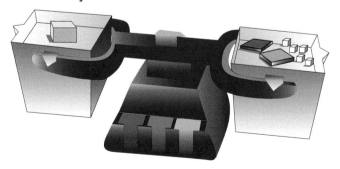

### Tip

A scale is the device that is used to measure mass. The common metric unit for measuring mass is the gram (g). To measure larger objects, a kilogram (kg) is used as the measurement unit. One kilogram is 1000 g. To measure smaller objects, a milligram (mg) is used. A milligram is $\frac{1}{1000}$ g. It is also useful to know that 1 mL of water has a mass of 1 g.

**You Will Need**
- small dry leaf
- apple
- scales

1. Examine a small dry leaf and an apple.

2. Explain which metric unit you would use to measure the mass of each object.

3. Estimate the mass of each object and record your estimate.

4. Measure the mass of each object. Record your measurements. How do they compare with your estimates?

5. Convert the mass of each object you measured into one other mass measurement unit. For example, if you measured the mass of an apple in grams, express it in milligrams.

### Vocabulary

**gram (g):** A metric unit used to measure mass
**kilogram (kg):** A metric unit, which is 1000 g
**mass:** The amount of matter in an object; usually measured in grams, milligrams, or kilograms
**milligram (mg):** A metric unit, which is $\frac{1}{1000}$ g
**scale:** A device used for measuring mass

### Tip

Use what you already know when making estimates. For example, the mass of a full juice box feels similar to the mass of an apple.

## Build Your Understanding

### Estimate and Measure Mass

You Will Need
- scales
- various classroom objects

1. Work in a small group. Copy this chart into your notebook.

| Object | Estimate | Mass |
|---|---|---|
|  |  |  |
|  |  |  |

2. Estimate the mass of each object and record your estimates on the chart.

3. Measure the masses of five objects in your classroom. At least one object should be measured in milligrams, one in grams, and one in kilograms. Record the names of the objects in the first chart column.

4. Use the most appropriate scale to measure each object. How do your estimates compare with your measurements?

5. Convert each measurement to at least one other metric unit. Convert at least one of the five measurements to a larger unit and another to a smaller unit.

## What Did You Learn?

1. Meet with a classmate. Discuss and summarize the relationships among grams, milligrams, and kilograms.

2. Explain the method you would use to convert a measurement from grams to milligrams.

3. Explain the method you would use to convert a measurement from grams to kilograms.

## Practice

1. Copy and complete this measurement conversion chart in your notebook.

| Kilograms (kg) | Grams (g) | Milligrams (mg) |
|---|---|---|
| 1 kg | | |
| | 100 g | |
| | | 10 000 mg |

2. What is the best unit of mass to use when measuring each of the following objects? Why?
   a) an adult
   b) grapes in a bunch
   c) a scooter
   d) a slice of bread

In your notebook, fill in the blanks with the correct anwers.

**3.** There are ■ g in a kg.

**4.** There are 1000 mg in a ■.

**5.** There are ■ mg in a kg.

**6.** Choose five objects at home or in your classroom. Estimate their mass in g and kg. Now measure the mass of each object. How did the actual measurements compare with your predictions?

## A Math Problem to Solve

**7.** Imagine that the total mass of luggage allowed per person on a small plane is 50 kg. What are two possible luggage combinations that would be under the plane's mass limit? Show your answer using numbers, pictures, and words. Describe the strategy you used to solve this problem.

Luggage A = 22 kg        Luggage B = 17 kg

Luggage C = 32 kg        Luggage D = 27 kg

Luggage E = 8 kg

## Extension

Here are some other units of mass and their equivalents:

1 hectogram = 100 grams

1 decagram = 10 grams

1 centigram = 0.01 gram

1 decigram = 0.1 gram

**8.** Meet with a classmate. Take turns explaining how you would convert 1 g into each unit of measurement.

**9.** What pattern or patterns do you see that might help you make conversions among units of mass?

## Show What You Know

**1.** Imagine that you work for a company that makes containers for computer parts. Your job is to design containers made of plastic foam to protect each computer part. The plastic foam container must be 2 cm larger in length, 2 cm larger in width, and 2 cm larger in height than the computer part it will protect.

Here are the computer part dimensions:
- CPU: Length = 42 cm, Width = 20 cm, Height = 47 cm
- Monitor: Length = 37 cm, Width = 30 cm, Height = 40 cm
- Speaker: Length = 14 cm, Width = 10 cm, Height = 23 cm
- Keyboard: Length = 45 cm, Width = 15 cm, Height = 3 cm
- Printer: Length = 37 cm, Width = 15 cm, Height = 17 cm

**a)** Calculate the volume of each computer part.

**b)** What will the capacity be of each plastic foam container? Show your work.

**2.** Estimate which mass measurement below might match the computer part listed in question 1. Explain your estimates.

1 kg

10 kg

8 kg

5 kg

3 kg

**3.** It costs $0.10 per 100 g to ship each computer part. How much will it cost in total to ship all the computer parts based on your estimated masses?

## Lesson 9

# Exploring Coordinate Grids

FLIGHT PLAN

PLAN:
You will learn about a coordinate grid and the important features of the grid. As well, you will create your own coordinate grid and plot points on it.

DESCRIPTION:
Pilots need good maps to navigate, especially when there are few landmarks to guide them. Coordinates on maps allow bush pilots to pinpoint particular locations.

## Get Started

You can use a map to find a city or town. If you are unsure where the community is, you can use the map index to find its coordinates. Then, you follow lines from the coordinates on the top and side of the map to the area where the lines meet to find the city or town.

A coordinate grid is used to show positions. It consists of two number or letter lines, one placed across the other so that the two lines are perpendicular. The horizontal line is called the *x*-axis. The vertical line is called the *y*-axis. The point where the two axes meet is called the origin.

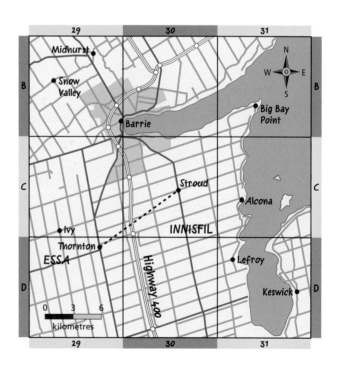

Two numbers called coordinates describe a position in the coordinate grid. To get to a position on the coordinate grid shown below, first move right of the origin along the *x*-axis, and then move up along the *y*-axis. For example, to get to point A, move three squares right along the *x*-axis, then two squares up along the *y*-axis.

The position of A is written as (3, 2).

*x*-coordinate    *y*-coordinate

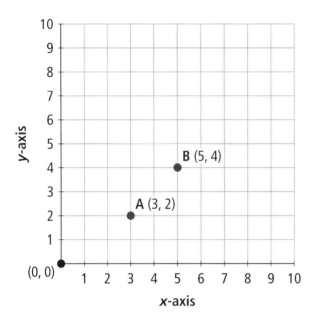

### Vocabulary

**coordinates:** An ordered pair of numbers used to describe a location on a grid or plane; for example, the coordinates (2, 3) describe a location on a grid found by moving 2 units horizontally from the origin (0, 0) and then 3 units vertically

**coordinate grid:** A grid that has data points named as ordered pairs of numbers, such as (5, 2)

**ordered pair:** A pair of numbers that form a location point on a coordinate grid; for example, (0, 2), (3, 4), (4, 5)

**origin:** In a coordinate system, the point where the *x*-axis and the *y*-axis intersect; it has the coordinates (0, 0)

**plane:** A flat or level surface that extends endlessly in all directions

**x-axis:** The axis that runs horizontally on a grid

**x-coordinate:** The first number in an ordered pair; it tells the distance to move right or left from (0, 0)

**y-axis:** The axis that runs vertically on a grid

**y-coordinate:** The second number in an ordered pair; it tells the distance to move up or down from (0, 0)

The coordinates are called an ordered pair, and they are written within parentheses ( ). It is very important that the coordinates are read and written in the correct order. Find the location of (2, 3). Notice that it is an entirely different location than point A (3, 2).

Remember that the first number in the pair tells the horizontal distance of the point from the origin. The second number tells the vertical distance from the origin. The origin has the coordinates (0, 0).

Work with a partner.

**1.** Give the coordinates of point B on the grid on page 289.

**2.** Find the location of (5, 2) on the grid. Then, find the location of (2, 5). Explain to your partner how the locations differ.

**3.** On a piece of grid paper, make your own coordinate grid. Follow the model in the illustration and label important features such as the origin and *x*- and *y*-axes. Place three different lettered points on the grid. On a separate piece of paper, create a key in which you record the coordinates for each point.

**4.** Exchange your grid with your partner. Challenge him or her to identify the coordinates for each point.

**Journal**

Imagine you have to explain how to read and plot coordinates to someone who is not familiar with them. Write simple, clear, step-by-step instructions. You will need to draw a labelled picture and include examples. Make sure to explain all mathematics terms.

## Build Your Understanding

### Locate and Plot Points on a Coordinate Grid

You Will Need
• grid paper
• ruler

**1.** Record the coordinates for each point on this coordinate grid.

**2.** On grid paper draw a coordinate grid. Label important features of the coordinate grid that help you to locate a position.

**3.** Place points at the following coordinates on your grid:
   **a)** A (6, 2)   **b)** B (3, 3)
   **c)** C (10, 4)   **d)** D (7, 1)
   **e)** E (2, 6)

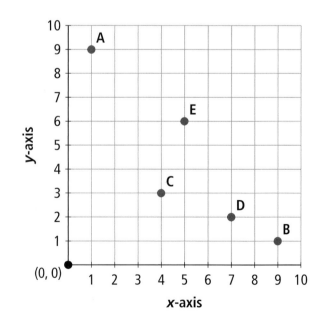

**4.** Draw another coordinate grid. Make it into a map of your classroom by drawing symbols at particular locations; for example your desk, the teacher's desk, a bookshelf, and so on. Create an index in which you give coordinates to help someone find the exact location of important classroom features.

**5.** Exchange your map with a classmate and challenge him or her to find the classroom positions using coordinates.

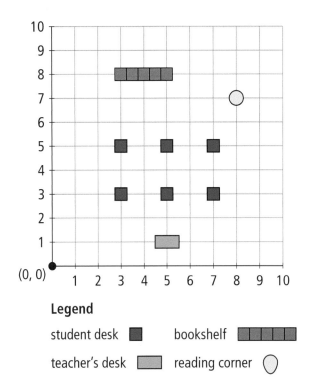

**Legend**

student desk ■    bookshelf ▨▨▨▨▨

teacher's desk ▭    reading corner ◯

## What Did You Learn?

**1.** When locating the point at (3, 7), which way would you move first, to the 3 or the 7? Why?

**2.** Explain to a classmate how the coordinates (2, 7) and (7, 2) differ.

**3.** Which of these ordered pairs is the coordinates of location Z in the grid?

   **a)** (3, 8)      **b)** (2, 7)

   **c)** (9, 1)      **d)** (8, 3)

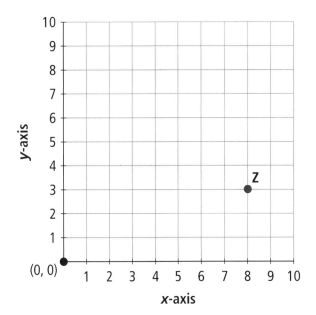

## Practice

**1.** Draw a coordinate grid in your notebook. Plot these points on your grid: (3, 5), (3, 11), (8, 5), (8, 11).

**2.** Join the points you have plotted. What shape have you made?

**3.** Plot points to make the following shapes:

    **a)** equilateral triangle    **b)** rectangle    **c)** parallelogram

    **d)** hexagon    **e)** rhombus    **f)** pentagon

**4.** Record the coordinates that are the vertices of each shape.

## A Math Game to Play

You Will Need
- markers (yellow, green, red)
- blank coordinate grids

### How to Play

Work with a partner.

**1.** Each of you needs your own blank coordinate grid that your partner cannot see. Decide who will be Partner A and who will be Partner B.

**2.** Partner A can use a yellow marker to plot the position of five treasures on his or her map at any coordinates he or she chooses.

**3.** Partner B can try to find Partner A's coordinates by guessing ordered pairs. (Remember, Partner B cannot see Partner A's grid.)

**4.** When Partner B guesses correctly, he or she can mark the points on his or her own grid with a green marker. When Partner B guesses incorrectly, he or she can mark the misses with a red marker.

**5.** When Partner B guesses all of Partner A's hidden treasure, both partners can switch roles.

### Show What You Know

Review: Lesson 9, Coordinate Grids

**1.** Use a coordinate grid to plot eight points that make an interesting shape.

**2.** Record the ordered pairs of each plotted point.

**3.** Give your list of ordered pairs to a classmate, and have him or her try and create the same interesting shape that you did. Are your shapes the same? Why or why not?

# Chapter Review

**1. a)** Estimate the percent of the grid that is red, green, blue, and unshaded.

**b)** Now calculate the percent of each section to check your estimates.

**c)** Express each percent as a fraction, a decimal, and a ratio.

**2.** Answer the following questions:

**a)** 456
234
712
+ 981

**b)** 45 678
− 9 234

**c)** 456
x 34

**d)** 27)‾2876

**e)** 512
290
312
+ 856

**f)** 35 312
− 4 035

**g)** 298
x 32

**h)** 34)‾7076

**3.** In your notebook, draw nets for the following prisms:

**a)** a rectangular prism

**b)** a cube

**c)** a triangular prism

**d)** a hexagonal prism

**e)** an octagonal prism

**4. a)** Copy and complete the chart shown below in your notebook:

| Pyramid | Number of Vertices | Number of Edges | Number of Faces |
|---|---|---|---|
| square-based | | | |
| triangular-based | | | |
| pentagonal-based | | | |
| octagonal-based | | | |

**b)** In your notebook, draw a net for each pyramid in the chart.

**c)** Design as many different nets for a triangular pyramid as you can, using isometric dot paper.

**5.** Estimate and then calculate the volume of the figures shown below. Each cube represents 1 cm³.

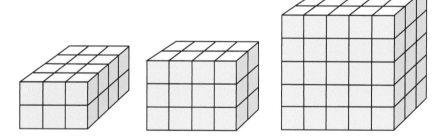

**6.** What is the volume of the figure shown below? Express your answer in the following units:

**a)** cubic centimetres

**b)** cubic millimetres

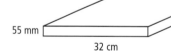

1 m

55 mm

32 cm

**7.** What is the capacity of this container? Express your answer in the following units:

**a)** cubic centimetres

**b)** millilitres

**c)** litres

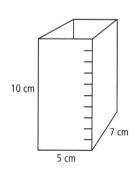

10 cm

7 cm

5 cm

**8.** List the following from smallest to largest according to their capacity:
- bucket
- tablespoon
- pop can
- eyedropper
- bowl
- juice jug

 **9.** Copy and complete this chart in your notebook.

| Object Number | Mass in mg | Mass in g | Mass in kg |
|---|---|---|---|
| 1 | 4675 | | |
| 2 | | 156 | |
| 3 | | | 0.6 |

**10.** Look at this coordinate grid and answer the following questions:

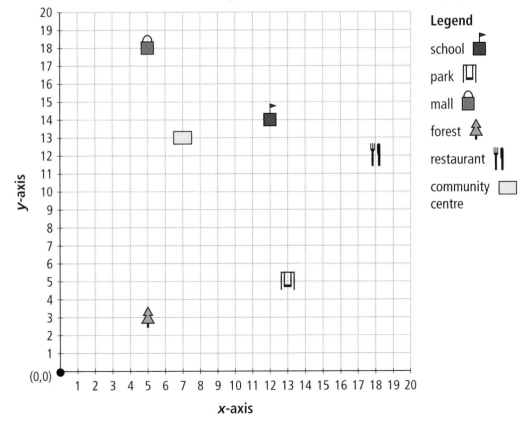

**Legend**

school ▣

park ⎁

mall ▢

forest 🌲

restaurant 🍴

community centre ▢

 **a)** Record the location of each landmark in the legend.

**b)** Following the grid coordinates, explain how you would get from the school to the forest.

# Chapter Wrap-Up

**Chapter 6**

You have reached the end of this chapter. Now it is time to apply what you have learned to complete your project!

**You Will Need**
- scissors
- tape
- string
- an electric fan
- stiff wire
- block of wood

Work in teams of four or five. You have learned about different three-dimensional shapes such as prisms, pyramids, and cubes. Each group will select a different polyhedron, create a net for the solid, and use the net to create the shape. One group will create an airfoil as shown below.

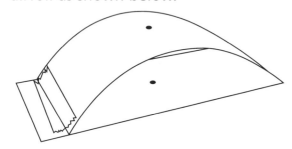

1. Which three-dimensional shape do you think will make the best wing? Record your prediction, and create a net for your shape.

2. Fold and tape your net to construct the three-dimensional shape.

3. Make two holes in the solid or airfoil (depending on which you have) as shown in the illustration.

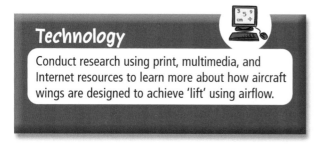

**Technology**

Conduct research using print, multimedia, and Internet resources to learn more about how aircraft wings are designed to achieve 'lift' using airflow.

**4.** Place the solid or airfoil on a wire in front of a fan as shown in the diagram below. Can you achieve lift with your shape? If so, time how long your wing shape can fly.

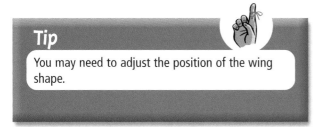

**Tip**

You may need to adjust the position of the wing shape.

**5.** Graph the class results.

**6.** What can you conclude from your graph about the best wing shape? How do the results compare with your predictions? How did the surface of the shape affect its ability to achieve and maintain lift?

**7.** Add objects such as feathers or paper clips that will change the mass of your wing shape. How does the increased mass affect the ability of your wing shape to fly?

**8.** Meet as a class to discuss your results in terms of lift, thrust, drag, and gravity.

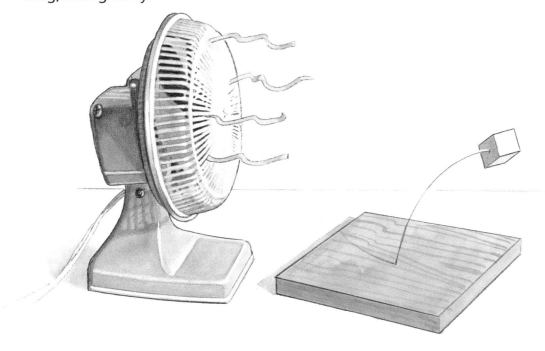

# Problems to Solve

Here are some fun problems for you to solve. For each problem, a helpful problem-solving strategy is included for you to use. For some problems, you will have the chance to choose the strategies you want to use.

## Problem 11

# Treasure Hunt

*STRATEGY: WRITE AN EQUATION*

*Sometimes, coming up with a math equation can help you solve a problem.*

*OBJECTIVE:*

*Demonstrate an understanding of whole numbers and place value*

## Problem

A treasure hunter in England used a metal detector to search for hidden treasure in farm fields. On the third day of searching, he found $234.00 worth of silver coins. On the fourth day, he found 2.5 times the amount found on the third day. On the fifth day, he found $578.00 worth of gold, which was double the amount found on the second day. What was the total worth of the treasure found over the 5 days, if the treasure hunter found the most treasure on the first day? Give one possible answer.

### Tip

**Problem-Solving Steps**
1. Understand the problem
2. Pick a strategy ("Write an Equation")
3. Solve the problem
4. Share and reflect

## Reflection

1. What information did you have about the problem when you began to solve it?

2. What did you need to figure out?

3. What type of calculations did you need to do?

4. How can you check your answer?

5. Why might your answer be different from another classmate's?

## Extension

1. While doing research on the Internet, you came across the following information on personal fortunes:

   Mr. Payday has a net worth of $45 678 456.00.

   Ms. Lotto has a net worth of $56 771 895.00.

   Mr. Moneybags has a net worth of $67 982 890.00.

   Place the numbers on a place-value chart. What patterns do you notice?

2. Write your own problem using large numbers.

**Problem 12**

# Organizing the Student Council

STRATEGY: USE LOGICAL REASONING

Logical reasoning helps you think about and solve the problem using information that you know.

OBJECTIVE:

Make conclusions from data

## Problem

**You Will Need**
• 1-cm grid paper

The student council works with your principal and teachers to make decisions that affect you. Imagine that four of your classmates (Ajay, Joey, Li, and Claire) make up the student council in your school. The possible jobs in the student council are President, Vice President, Treasurer, and Secretary. Assign one job to each student.

**1.** At a student council meeting taking place in a classsroom, the President is seated beside the Vice President. The Vice President is sitting 3 seats away from the Secretary, in a diagonal direction. The Treasurer is sitting beside the Secretary. Where are the 4 student council members sitting in relation to one another? Using a grid to help you, give one possible answer.

**2.** Change the jobs around, so each student has a different job. Repeat question 1.

## Reflection

**1.** What did you know about the problem before you started to solve it?

**2.** What did you need to figure out?

**3.** Look at your answers. What do you notice about them? Why does this make sense?

**4.** Share your results with two classmates. Why do you think your answers might be different?

## Extension

Rewrite question 1 of the Problem, but this time make up new clues. Share your question with another student.

**NUMBER SENSE AND NUMERATION**

Problem 13

# Temperature Change

STRATEGY: YOUR CHOICE

OBJECTIVE:

Perform calculations involving integers

## Problem

The temperature in Calgary can change quickly, especially in the wintertime. Chinooks (warm west winds) can make a negative temperature positive in a short period of time. For example, look on page 302 at the temperature chart.

**1.** How many degrees did the temperature change from Monday P.M. to Tuesday P.M.?

**2.** Imagine that the temperature rises again to a positive value on Thursday. The sum of the A.M. temperature and the P.M. temperature equals +16. What might the A.M. and P.M. temperatures be for Thursday?

| January 13-17 | | | | | |
|---|---|---|---|---|---|
| | Monday | Tuesday | Wednesday | Thursday | Friday |
| A.M. Temp. | –13°C | +6°C | 0°C | +?°C | ?°C |
| P.M. Temp. | –18°C | +8°C | –1°C | +?°C | ?°C |

**3.** If the temperatures on Friday dropped from Thursday by $\frac{1}{3}$, both in the A.M. and P.M., what would Friday's temperatures in the morning and the afternoon be?

## Reflection

**1.** What did you know about the problem before you began to solve it?

**2.** What did you need to figure out?

**3.** Do you think the strategy you chose was a good strategy to use? Why?

## Extension

**1.** What is the difference between Tuesday's afternoon temperature and Thursday's afternoon temperature?

**2.** How much colder is it on Monday than on Thursday? Give an answer for both A.M. and P.M.

**3.** In what other situations would you find negative numbers?

**4.** Make up your own integer problem.

## Problem 14

# A Very Old Tree

*STRATEGY: YOUR CHOICE*

*OBJECTIVE:*
*Use division skills*

## Problem

One of the oldest living trees in the world is called "Methuselah." It is a 4700-year-old bristlecone pine growing in California in the United States.

1. How many millenniums does the age of this tree cover?

2. How many centuries does it cover?

3. How many decades does it cover?

4. Canada became an official country in 1867. What math problem could you write comparing Canada and Methuselah?

## Reflection

1. What information did you have about the problems before you began to solve them?

2. What did you need to figure out?

3. Which strategy did you pick? Why?

4. How do you know your answers make sense?

5. Have a classmate do the problem you created for question 4.

## Extension

**You Will Need**
• calculator

1. One of the oldest living people is 121 years old. How does the age of Methuselah compare to this person's age?

2. How does the age of this tree compare to the combined age of all of your classmates? What do you think is the best way to find the solution? You may use a calculator.

## Problem 15

# Bird Speeds

*STRATEGY: YOUR CHOICE*

*OBJECTIVE:*

*Perform calculations involving measurement and division; write and compare ratios*

## Problem

1. Scientists have discovered that when female peregrine falcons dive downwards at their prey, they can reach speeds of over 300 km/h. If an eagle dives at 150 km/h and a turkey vulture dives at 100 km/h, how can you compare the speed of the 3 creatures using ratios?

2. Make a chart showing km/h and m/s for each bird.

3. Comment on any patterns that you notice.

1. What did you know about the problem before you started to solve it?

2. What did you need to figure out?

3. Which strategy did you use? Why?

4. List the steps you followed to solve this problem.

5. Compare your answer with another classmate's. Did you both follow the same steps? Explain.

## Extension

You Will Need
• pylons
• tape measure
• stopwatch

1. How does the speed of a female peregrine falcon compare to the speed of an Olympic sprinter who can run 100 m in 10 s?

2. Go outside and use pylons and a tape measure to mark the distance a falcon can fly in 1 s. Using a stopwatch, time how long it takes you to run this distance. Write a ratio to show the comparison.

# Unit 3
## The Math of Starting a Business

It's time to start your own business! Math is an important part of business.

In Chapter 7, you will work with large numbers, rate and ratio, percent and fractions. You will plot coordinates and learn about prime and composite numbers and multiples and factors. At the end of the chapter you will present a business plan to a group of Grade 5 students.

In Chapter 8, you will conduct surveys, display data, explore probability, and solve problems as you examine the contents of flyers. At the end of the chapter you will create your own flyers.

In Chapter 9, you will explore area, volume, mass, patterns, coordinates, transformations, and tessellations as you decide how you want to spend your hard-earned money. At the end of the chapter, you will design CD jackets, as well as containers in which to ship them.

Have fun and good luck!

# Chapter 7

# The Numbers of Business

In this chapter, you will develop your math skills as you start your own business delivering flyers to homes in your neighbourhood.

In this chapter, you will
- practise estimating large numbers
- work with ratios and rates
- calculate percents and fractions
- create number lines
- plot coordinates
- practice multiplication and division rules
- explore prime and composite numbers

- explore least and greatest multiples and factors

At the end of this chapter, you will create a presentation for Grade 5 students about organizing a flyer delivery business.

Here are some questions to start you thinking:

1. Where do you see percents and fractions used in everyday life?

2. What do you need to know in order to read a map?

## Lesson 1

# Making Estimates

*BUSINESS PLAN*

*PLAN:*
You will estimate how much time it takes to prepare your flyers for delivery.

*DESCRIPTION:*
You have been hired by a marketing company to deliver flyers to homes in your neighbourhood every Saturday morning. Your first job is to organize the flyers into bundles.

## Get Started

Here are two common strategies you can use to estimate.

### Method 1
When you work with numbers that have the same number of digits, you can cluster them to the same value.

3100 ⟶ 3000
2900 ⟶ 3000
3050 ⟶ 3000
2950 ⟶ 3000

### Method 2
When you work with numbers with a different number of digits, you can round them to the nearest ten, hundred, or thousand.
These numbers have been rounded to the nearest hundred.

580 ⟶ 600
425 ⟶ 400
689 ⟶ 700
1694 ⟶ 1700

### Vocabulary

**clustering:** A rule used to change the value of like numbers to the closest common place value
**rounding (up or down):** A rule used to make a number approximate to a certain place value.

Use clustering or rounding to estimate the answers to these equations:

**1.** 479 + 590 + 630 =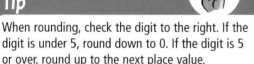

**2.** 202 + 197 + 210 = ▣

**3.** 640 + 340 + 700 = ▣

**4.** 57 x 52 = ▣

**5.** 55 + 62 + 8 + 9 = ▣

**6.** 1002 + 46 + 63 = ▣

**7.** 1.09 + 0.88 = ▣

**8.** 6.8 + 0.2 + 7.9 = ▣

**9.** 0.002 + 0.009 + 0.010 = ▣

**10.** 26 x 21 = ▣

### Tip

When rounding, check the digit to the right. If the digit is under 5, round down to 0. If the digit is 5 or over, round up to the next place value.

## Build Your Understanding

### Estimate

You Will Need
- seven stacks of paper
- stopwatch or watch with a seconds hand

There are 7 stacks of flyers. There are 770 flyers in total. You must make bundles of flyers each containing 1 paper from each of 7 stacks.

Work with a classmate.

**1.** How many bundles of flyers will you make.

**2.** Estimate how much time it will take you to prepare all of your bundles. Record your estimate.

**3.** Take turns with your partner making 1 bundle with 1 paper from each of the 7 stacks. Time how long it takes each of you to make 1 bundle.

**4.** Based on the time it took you to make 1 bundle, calculate how much time you would need to bundle all 770 flyers. Compare your calculation with your estimate in question 2.

## What Did You Learn?

**1.** How did you estimate your time?

**2.** How close was your estimate to your recorded time?

**3.** What operations did you use in this activity?

## Practice

Solve these equations:

**1.** $420 + 500 + 905 + 254 = \blacksquare$

**2.** $690 + 710 + 605 + 590 = \blacksquare$

**3.** $785.0 + 1.2 + 700.0 + 199.0 = \blacksquare$

**4.** $12\ 944 - 7865 = \blacksquare$

**5.** $20\ 009 - 4750 = \blacksquare$

**6.** $16\ 498 - 6100 = \blacksquare$

**7.** Check your answers to questions 1 to 6 by adding or subtracting.

**8.** What strategies did you use to solve questions 1 to 6? Explain why you chose these strategies.

Solve these equations:

**9.** $360 \times 12 = \blacksquare$      **10.** $489 \times 11 = \blacksquare$

**11.** $625 \times 19 = \blacksquare$      **12.** $850 \times 50 = \blacksquare$

**13.** $500 \div 10 = \blacksquare$      **14.** $600 \div 5 = \blacksquare$

**15.** $7250 \div 25 = \blacksquare$      **16.** $900 \div 90 = \blacksquare$

**17.** $1001 \div 10 = \blacksquare$      **18.** $750 \div 50 = \blacksquare$

**19.** Check your results by multiplying or dividing.

**20.** What strategies did you use to solve questions 9 to 18? Explain why you chose these strategies.

**21.** Describe a situation in which clustering would be a good estimation strategy.

**22.** Describe a situation in which rounding would be a good estimation strategy.

## A Math Problem to Solve

**23. a)** It took Eric 1 h 15 min to make all his bundles. It took him 15 s to make each bundle. How many bundles did Eric make? Show your solution using numbers, pictures, and words.

  **b)** Explain to a classmate the strategy you used to solve the problem.

## Extension

**24.** When you estimate a number that is lower than the actual number, you underestimate. When you estimate a number that is higher than the actual number, you overestimate. Usually, it is better to overestimate than to underestimate.

Give two reasons why you agree or disagree with this statement.

**Lesson 2**

# Exploring Ratios

*BUSINESS PLAN*

*PLAN:*
You will work out ratios to compare your route to the length of your classmates' routes.

*DESCRIPTION:*
Several of your classmates have a flyer route like yours. Some routes are longer than yours, some are shorter.

**Legend**

route 1 ——    route 4 ——
route 2 ——    route 5 ——
route 3 ——

N W E S

Walnut St.

Centre Ave.

Maple Ave.

Acorn Ave.

King St.  Wellington St.  Main St.  Queen St.  Oak St.

## Get Started

You play hockey with some of your classmates. The ratio of players to referee is 10 players to 1 referee. You can write this three ways:

10 to 1       10:1       $\frac{10}{1}$

Ratios are considered to be equivalent if they can be made the same by multiplying or dividing the terms of the ratio by the same number. The ratios $\frac{10}{1}$, $\frac{100}{10}$, and $\frac{20}{2}$ are examples of three equivalent ratios.

$\frac{10}{1}$    $\frac{100 \div 10}{10 \div 10} = \frac{10}{1}$    $\frac{20 \div 2}{2 \div 2} = \frac{10}{1}$

1. Write an equivalent ratio for these examples:
   a) 12 books to 10 students
   b) 25 juice boxes to 5 students
   c) 10 papers to 8 magazines
   d) 9 pens to 3 pads of paper

In question 1, you worked with ratios that compared part to part. Read these statements:

On the street where you make most deliveries, there are 3 elm trees, 2 birch trees, and 5 trees in total.

In this statement, the 5 trees are the whole and the elm and birch trees are part of the whole.

**2.** Use the above statements to write ratios for the following examples. Specify whether you are comparing part to part, part to whole, or whole to part.

   **a)** elm trees to birch trees

   **b)** all trees to birch trees

   **c)** elm trees to all trees

   **d)** birch trees to elm trees

> ## Vocabulary
>
> **equivalent ratios:** Ratios that represent the same fractional number or comparison; for example, 1:2, 2:4, 3:6.
> **ratio:** A comparison of two numbers with the same unit. For example, 1:4 means "1 compared with 4" and can be written as a fraction: $\frac{1}{4}$.

## Build Your Understanding

### Write Ratios

This chart shows the number of houses on each student's delivery route.

**1.** Write a ratio that compares the number of houses on Ruth's route to the number on Phillip's route.

**2.** Write a ratio that is equivalent to the ratio in question 1.

**3.** An equivalent ratio for the number of houses on two students' routes may be represented as $\frac{3}{2}$. Which routes are associated with this ratio?

| Name | Number of Houses on Route |
|------|---------------------------|
| Ruth | 110 houses |
| Tony | 120 houses |
| Eric | 80 houses |
| Phillip | 60 houses |
| Mahalia | 40 houses |

**4.** Write a ratio that compares the total number of houses on all the routes to the number of houses on Tony's route.

**5.** Reverse the numbers in question 4, and identify whether you are comparing part to part, part to whole, or whole to part.

## What Did You Learn?

Describe how learning about ratios can help you understand comparisons.

## Practice

Draw a picture to show each of the following ratios. The first one has been done for you.

**1.** 2 students to 5 bicycles

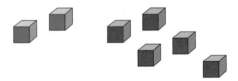

**2.** 8 players to 8 uniforms

**3.** 1 teacher to 3 students

**4.** $\frac{14}{5}$

**5.** 7:6

**6.** $\frac{6}{10}$

**7.** Create a table with at least six columns. In the first column, write a ratio in fraction form. In the remaining columns, write equivalent ratios for the first ratio, as shown in this example.

| $\frac{1}{3}$ | $\frac{2}{6}$ | $\frac{3}{9}$ | $\frac{4}{12}$ | $\frac{5}{15}$ | $\frac{6}{18}$ |
|---|---|---|---|---|---|
| | | | | | |
| | | | | | |
| | | | | | |
| | | | | | |

Write an equivalent ratio for each of the following:

**8.** $\frac{50}{1}$

**9.** $\frac{20}{4}$

**10.** 16 to 1

**11.** $\frac{100}{50}$

**12.** 6:72

**13.** $\frac{106}{2}$

**14.** 80:8

**15.** $\frac{1.0}{0.1}$

**16.** Use your classroom to write ratios for the following examples. Specify whether you are comparing part to part, part to whole, or whole to part.

    **a)** girls to boys         **b)** all students to girls

    **c)** boys to all students     **d)** boys to girls

**17.** Write your ratios from question 16 in fraction form.

## Extension

**18.** Use the numbers in Build Your Understanding to write three ratio questions for a classmate to solve. Your questions can be word or number sentences. If needed, give a clue for each question. Check your classmate's answers, and compare methods for solving problems.

**19.** Write ratios that compare things in your classroom. Share your ratios with the class.

**20.** Write your ratios from question 18 in fraction form. Write three equivalent ratios for each ratio. Explain how you know they are equivalent to the first ratio.

## Lesson 3

# Working With Rates

*BUSINESS PLAN*

*PLAN:*
*You will work with rates to find your earnings on a weekly, monthly, and yearly basis and you will explore unit rates.*

*DESCRIPTION:*
*It's time to think about what you might earn each week, each month, and each year.*

## Get Started

You will be working with decimals. Here are some practice equations to get you started.

**1.** $1.09 \div 7 = $ ▮  **2.** $6.43 \div 2 = $ ▮  **3.** $0.41 \div 10 = $ ▮

**4.** $2.7 \div 8 = $ ▮  **5.** $4.77 \div 3 = $ ▮  **6.** $9.9 \div 3 = $ ▮

## Build Your Understanding

### Calculate Rates

Imagine that you will be paid $0.10 for each of the 110 bundles of flyers that you deliver each week.

**1.** Calculate and record how much you will make in

    **a)** 1 week    **b)** 1 month    **c)** 1 year

**2.** With a partner, create a list of other examples of rates found in everyday life.

## Calculate Unit Rates

Review the definitions of rate and unit rate. Explain the difference between rate and unit rate in your own words.

You are hungry as you set out to deliver your flyers. You buy 3 chocolate wafers to keep you going until you reach home. The price of the wafers is 3 for $0.99. Here is how you can figure out the rate for each wafer:

Rate: $0.99 for 3 wafers $\frac{0.99}{3} = 0.33$ Each wafer costs $0.33.

1. Find the unit rate for each of the following:
   a) 2 hats for $15.00
   b) 3 books for $4.98
   c) 5 magic pens for $10.00
   d) 2 pairs of sandals for $20.00
   e) 3 watches for $24.99
   f) 6 pairs of socks for $12.96
   g) 2 CDs for $21.98
   h) 4 art kits for $39.96

2. Decide which of the items in question 1 you will buy for your friend's birthday. Try to include as many items as you can to a maximum of $40.00. You can buy only 1 of each item.

## What Did You Learn?

1. Explain how to change a rate into a unit rate. Make up a question to show your method.

2. With a partner, brainstorm situations in everyday life where you often see numbers shown as rates.

**Vocabulary**

**rate:** A ratio that compares two quantities with different units of measurement; for example, price per kilogram

**unit rate:** A ratio that compares a number to 1; for example, the delivery fee you earn for delivering one set of flyers

Decide whether the following statements are true or false:

**1.** Kilometres per hour is an example of a unit rate.

**2.** The charge for gasoline per litre is an example of a rate.

**3.** Four CDs for $5.99 is an example of a rate.

**4.** The price of 3 L of milk is an example of a unit rate.

## A Math Problem to Solve

**5.** Using the rate given in Build Your Understanding, how much would you earn per year delivering flyers if, halfway through the year, you added 35 new houses to your route? Show your solution using numbers, pictures, and words. When you have finished, describe to a classmate how you found the answer to this question.

## Lesson 4

# Calculating Rates and Percents

*BUSINESS PLAN*

*PLAN:*
**You will calculate rates and percents based on an increase in your pay.**

*DESCRIPTION:*
*You will use the rate from Lesson 3, but in this lesson you are going to get a raise.*

## Get Started

Here is an example to get you started. Mahalia's sister, Margaret, makes $10.00 per hour as a store clerk. She gets a 10% pay increase.

Margaret's unit rate: $10.00 per hour

Pay increase: 10%

How much more per hour will Margaret make? How did you determine this?

What is Margaret's new unit rate?

Percentages can be shown three ways:

| As a fraction | As a decimal | As a percent |
|:---:|:---:|:---:|
| $\frac{10}{100}$ | 0.10 | 10% |

You could calculate Margaret's new unit rate as follows:

0.10 x 10 = 1

Therefore, Margaret will make $1.00 more per hour.

$10.00 + $1.00 = $11.00

Margaret will now make $11.00 per hour.

Write these percents as decimals and fractions:

**1.** 25%   **2.** 7%   **3.** 12%   **4.** 9%   **5.** 87%

**6.** 50%   **7.** 3%   **8.** 62%   **9.** 14%   **10.** 75%

## Build Your Understanding

### Calculate Rates and Percents

In Lesson 3, your rate was $0.10 per bundle delivered. Now imagine that you are going to receive a 20% raise. How much more will you make

1. each week?

2. each month?

3. each year?

4. How much more will you make if you received a 30% raise?

You may use a calculator to help you.

**Tip**

Use this example to help you. Joey makes $1200.00 per year delivering flyers. He received a 10% pay raise. To find out how much more he would make, Joey did this calculation:

0.10 x 1200 = 120

Joey will make $120.00 more each year.

## What Did You Learn?

1. What strategy did you use to answer the questions in Build Your Understanding? Compare your strategy to a classmate's. Were your answers the same?

2. Discuss with a classmate another way to find the answers.

## Practice

Use a calculator to help you answer the following questions. Show your work.

1. Ajay's mother makes $20 000.00 per year. She received an 8% raise. How much more will she make next year?   **8%**

2. Ahanu's brother makes $15 000.00 per year. He received a 5% raise. How much more will he make next year?   **5%**

3. Phillip's sister makes $8.00 per hour. She received a 5% pay raise. How much more will she make per hour?   **5%**

4. Tony's father earns $15 000.00 per year. His company decreased everyone's salary by 8%. How much less will he make next year?   **8%**

## Math Problems to Solve

**5.** Imagine you had a job delivering flyers and you were going to receive a 10% raise each year for 5 years. Starting with a rate of $0.10 per bundle, calculate your salary for each of the 5 years.

**6.** Look at these different types of graphs. Decide which type of graph would best display your answer to question 5. Create a graph to display your data. Compare your graph to a classmate's.

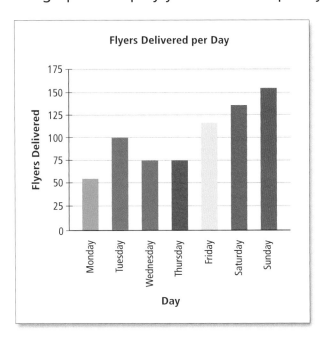

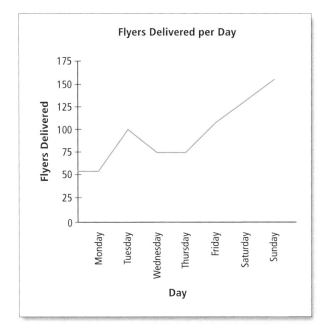

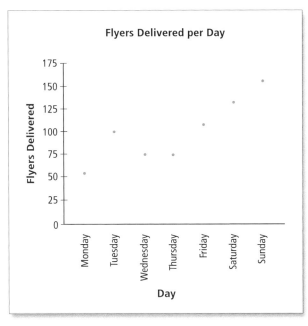

## Lesson 5

# Plotting on a Number Line

**BUSINESS PLAN**

*PLAN:*
You will plot wages on a number line.

*DESCRIPTION:*
It's time to explore number lines. You will do this by plotting whole numbers, fractions, and decimals.

## Get Started

You can use number lines to order and compare whole numbers, fractions, and decimals.

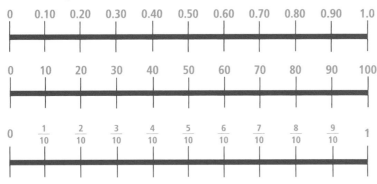

1. Order these fractions from least to greatest:

   **a)** $\dfrac{1}{10}$   $\dfrac{10}{10}$   $\dfrac{4}{10}$   $\dfrac{2}{10}$

   **b)** $\dfrac{93}{100}$   $\dfrac{47}{100}$   $\dfrac{53}{100}$   $\dfrac{1}{100}$

2. Order these decimals from least to greatest:

   **a)** 5.5   0.5   0.545   0.9

   **b)** 8.09   1.10   0.58   0.001

3. Write these decimals as percents:

   **a)** 0.02   **b)** 0.5   **c)** 0.09   **d)** 0.3

4. Write these fractions as percents:

   **a)** $\dfrac{7}{100}$   **b)** $\dfrac{3}{50}$   **c)** $\dfrac{19}{100}$

## Create Number Lines

Tony makes $1.00 for every 10 bundles of flyers he delivers, Mahalia is paid $0.40 per 10 bundles, Ajay receives $0.90 per 10 bundles, Phillip makes $0.20 per 10 bundles, Ahanu receives $0.80 per 10 bundles, and Joey earns $0.70 per 10 bundles.

1. Create a number line in your notebook like the one below, then plot how much each student is paid for every 10 bundles.

2. Convert the amounts from question 1 to fractions, then plot them on a number line like the one below.

3. Now convert the amounts from question 1 to percents, then plot them on a number line like the one below.

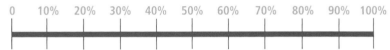

1. Compare your number lines with those of another group of students. What reason might there be for differences between the number lines?

2. **a)** What information do the number lines show?

   **b)** How else could you show the information that is on the number lines?

Plot these numbers on a number line:

**1.** $\frac{1}{6}$, $\frac{5}{6}$, $\frac{2}{8}$, $\frac{7}{8}$, $\frac{1}{9}$

**2.** 0.05, 0.1, 0.02, 0.4, 0.09

**3.** 12%, 6.5%, 3%, 4.5%, 9%

Convert these ratios to fractions:

**4.** 2:5     **5.** 4:5     **6.** 7:8     **7.** 2:12     **8.** 4:6

**9.** 5:4     **10.** 8:7     **11.** 1:12     **12.** 11:12     **13.** 7:9

Convert these ratios to decimals:

**14.** $\frac{67}{100}$     **15.** 3:2     **16.** $\frac{108}{100}$     **17.** $\frac{33}{100}$

**18.** 60:25     **19.** $\frac{14}{100}$     **20.** $\frac{0.001}{100}$     **21.** $\frac{0.001}{100}$

Convert these ratios to percents:

**22.** 4:5     **23.** 5:10     **24.** $\frac{76}{100}$     **25.** $\frac{0.04}{100}$

## A Math Problem to Solve

**26.** If you saved every cent you made from delivering flyers, how long would you have to work to buy a bicycle that costs $249.99? Use a rate of $0.10 per bundle and assume you deliver 110 per week. Use numbers, pictures, and words to solve this problem.

**Booklink**

**Fractals, Googols, and Other Mathematical Tales** by Theoni Pappas (Wide World Pub./Tetra: San Carlos, CA, 1993). Includes short stories and discussions that present such mathematical concepts as decimals, tangrams, number lines, and fractals.

## Show What You Know

1. Imagine that Joey is helping his older sister with her job at a sporting goods store. He has to put T-shirts on racks. It takes him approximately 90 seconds to take a T-shirt out of the box and put it on a clothes hanger. Estimate how long it would take him to put 200 T-shirts on the racks. Calculate. Did you overestimate or underestimate?

2. The store sells 2 pairs of shoes for every 1 T-shirt that it sells. Write the ratio in three different ways. If the store sold 300 pairs of shoes, how many T-shirts would they sell?

3. Imagine that a customer wants to spend $100.00. Here is the cost of the merchandise he wants:
   - running shoes $65.00
   - T-shirt $6.95
   - sweatshirt $29.99
   - baseball glove $34.99
   - tennis balls $4.59
   - baseball bat $45.65

   Including 7% sales tax, calculate the fewest items he could buy, then calculate the most items he could buy.

4. If the store decides to lower the prices by 20%, calculate how much the customer in question 3 would save.

5. Joey's sister is being paid $7.00 per hour. Then she receives a 15% pay raise.
   a) What would she make if she worked 15 hours per week?
   b) How much would she make in 2 months?
   c) How much would she make in a year?
   d) Explain in your journal how you solved this problem.

## Lesson 6

# Coordinate Grids (Part 1)

BUSINESS PLAN

PLAN:
**You will locate coordinates on a grid.**

DESCRIPTION:
You want to do your job as efficiently as you can. This means that you must plan a route so that you don't spend more time than necessary delivering your flyers.

## Get Started

**1.** State the coordinates of these points. The first one has been done for you.

  **a)** A = (3, 19)

  **b)** B = ■

  **c)** C = ■

  **d)** D = ■

  **e)** E = ■

**2.** Plot these points on the graph:

  F (6, 11)   G (15, 16)   H (1, 13)

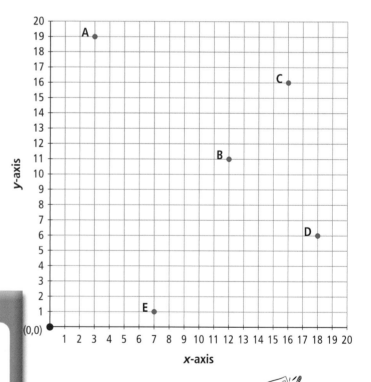

### Vocabulary

**coordinate grid:** A grid that has data points named as ordered pairs of numbers, such as (5, 2)
**ordered pair:** A pair of numbers that form a location point on a coordinate grid
**x-axis:** The axis that runs horizontally on a grid
**y-axis:** The axis that runs vertically on a grid

### Journal

Record in your journal why it is important to put the coordinates in the correct order when graphing points. Write a rule about the correct order.

## Locate Coordinates

You Will Need
• coordinate grid (provided below)

Each point represents one house on your route.

Work in small groups. Each student in your group will write the coordinates (locations) of each house. Compare your answers.

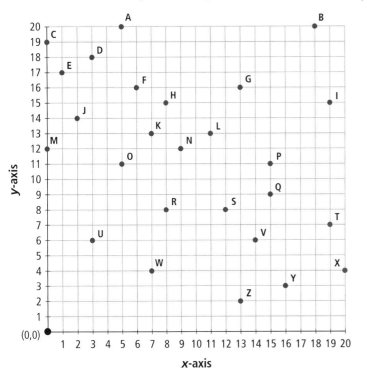

**1.** Describe why it is easy to locate a point using coordinates.

**2.** If your house is located at (0, 0), state the coordinates of the house closest to you. State the coordinates of the house farthest away.

**3.** If each line on the grid represents a street, list the coordinates of houses that are on the same street. What do you notice?

## Practice

You Will Need
• blank grid

Think about a mystery word, shape, or object.

**1.** Use a pencil to sketch your item lightly on the grid.

**2.** Plot points that will provide an outline of the item.

**3.** Record the coordinates for each point.

**4.** Erase your pencil marks.

**5.** Give your blank mystery grid and your list of coordinates to a classmate to plot the coordinates and solve the mystery.

## Lesson 7

# Coordinate Grids (Part 2)

*BUSINESS PLAN*

*PLAN:*
You will measure the length of your delivery route by locating points on a coordinate grid.

*DESCRIPTION:*
You have plotted your route using coordinates. Now it's time to figure out how far you will walk on your weekly route.

## Get Started

You Will Need
- metre stick
- stopwatch or watch with a second hand

Work with a partner.

In this activity, you will be working with a scale to measure distance.

**1.** Measure 100 m in the hallway or out in the schoolyard.

**2.** Time how long it takes you to walk 100 m.

**3.** Record your time.

## Calculate Distances

You Will Need
• copy of the map below

Use the scale on the map to calculate how long it would take you to deliver a set of flyers to each house on the grid below. You will begin and end at your house. You must stay on the roads (grid lines) and not cross diagonally over someone's lawn.

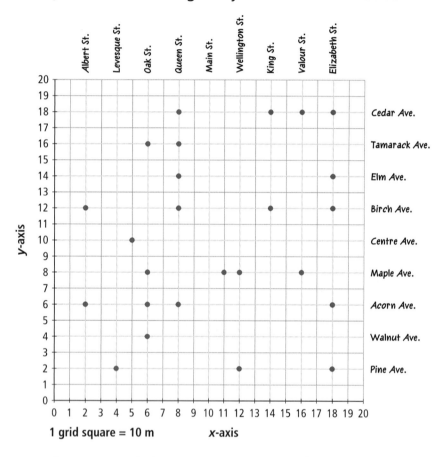

1 grid square = 10 m          *x*-axis

1. Choose a red dot to represent your house. What are its coordinates?

2. Create a delivery route and draw it on the copy of this map. Remember that you must start and finish at your house.

3. Calculate how many metres you would walk to deliver all the flyers.

4. Using your time data from Get Started, calculate the time it would take for you to deliver the flyers to all the houses.

## What Did You Learn?

1. Discuss with a classmate what you found most challenging about Build Your Understanding.

2. In a group of three or four students, compare the routes you created. How were they different?

## Practice

1. Use a provincial map to measure the distances between five towns or cities (including the place you live) and the province's capital.

2. Record each distance.

3. Look at the graph examples on page 321. Decide which of these graphs would best show your data.

4. Create a graph of your data.

5. Display your completed graph on a classroom bulletin board.

### Extension

6. Explore computer applications that help users build graphs. Use your knowledge to write step-by-step instructions to help someone learn how to build one type of graph using a computer application.

**Tip**

As you write your instructions, keep in mind that your readers may not have used this computer application before.

## Lesson 8

# Calculating Average Speed and Distance

*BUSINESS PLAN*

*PLAN:*
You will calculate your average speed and distance.

*DESCRIPTION:*
In the last lesson, you calculated how long it took you to deliver your flyers, based on the time it took you to walk 100 m. In this lesson, you will learn how to calculate average speeds.

## Get Started

1. Convert the following hours to minutes:

   **a)** 10 h    **b)** 15 h

   **c)** 6.5 h    **d)** 19.25 h

2. Convert the following minutes to seconds:

   **a)** 41 m    **b)** 12 m

   **c)** 23 m    **d)** 75 m

## Tip

Multiplication and division are handy operations to use when answering these questions.

3. Convert the following seconds to minutes:

   **a)** 1140 s    **b)** 480 s

   **c)** 1500 s    **d)** 240 s

## Build Your Understanding

### Calculate Speed

**You Will Need**
• stopwatch or a watch with a second hand

Work with a partner.

1. Have your partner measure how far you walk in 1 min.

**2.** Have your partner walk for 1 min, then measure the distance he or she has walked.

**3.** Do this activity five times and record the results.

Now you are going to find your average distance when walking for 1 min.

**4.** Add all of your measurements together.

**5.** Divide your total by the number of trials (5). The number you get is your average distance per 1 min.

**6.** Multiply by 60 to find your average walking speed per hour.

**7.** Round your speed to the nearest km/h, record how long it would take you to walk these distances:

**a)** 6 km          **b)** 0.5 km

**c)** 5.5 km        **d)** 21.5 km

**e)** 0.6 km        **f)** 7.4 km

> **Tip**
>
> Speed is a unit rate that shows the distance travelled in a single time period. For example, when 50 km are travelled in 1 h, the speed is 50 km per hour and is written 50 km/h.

**8.** For each speed and time given, calculate the distance travelled.

**a)** 7 km/h for 4 h          **b)** 85 km/h for 2 h

**c)** 70 km/h for 30 min      **d)** 100 km/h for 3.5 h

**e)** 800 km/h for 40 h       **f)** 500 km/h for 30 min

**g)** 1000 km/h for 10 h      **h)** 1500 km/h for 10.5 h

**i)** 12 km/h for 45 min

**9.** Calculate the speed needed in km/h to travel each distance in the given time. Use a calculator to check your answers.

**a)** 8 km in 4 h             **b)** 95 km in 30 min

**c)** 1000 km in 3 h          **d)** 500 m in 5 min

**e)** 750 m in 6 min          **f)** 550 km in 30 min

**g)** 10 000 km in 10.5 h     **h)** 20 m in 5 s

**i)** 140 m in 45 s

**10.** Create a graph or number line that shows the difference between your speeds for walking 100 m. Keep in mind that the objective of your visual aid is to show others your speed for walking these distances.

11. Present your visual aid to a small group of four or five students. Once everyone has shown his or her work, discuss similarities and differences.

### Booklink

**Math Mini Mysteries** by Sandra Markle (Macmillan Canada: Toronto, ON, 1993). From magic square puzzles to measuring wind speed, this book overflows with problems to investigate and solve.

## What Did You Learn?

1. Give reasons for your choice of visual aid to show the differences in your walking speeds.

2. Compare your work with that of other groups of students in the class. In what ways are they similar? In what ways are they different?

## Practice

### Math Problems to Solve

For each problem, show your solutions using numbers, pictures, and words. Use a calculator to check your work.

1. Li ran a 400 m track every day for a week. Her times were 53 s, 68 s, 49 s, 52 s, 47 s, 53 s, and 61 s. What was Li's average speed?

**2.** Jimmy drove 160 km to visit his sister Ruth. From there, he drove 210 km to visit his cousin. Finally, Jimmy drove 180 km to visit his grandmother. He didn't drive below 60 km/h or more than 80 km/h.

   **a)** What is the least number of hours that Jimmy might have driven on his trip?

   **b)** What is the greatest number of hours he might have driven?

**3.** Halle decides to test her new car on a road trip to visit her niece Claire. She drives for 8 h the first day and 3 h the next day. Her average speed on the first day is 80 km/h. Her average speed on the second day is 60 km/h. Find the average speed per hour over the two days that Halle drove.

**4.** Share with a partner the strategies you used to solve questions 1 to 3.

## Show What You Know

### Review: Lessons 6 to 8, Coordinate Grids and Averages

**1.** Place 4 short lines that will represent ships on a coordinate grid; one that is 2 points long, one that is 3 points long, one that is 4 points long, and is one that is 5 points long. Have your partner call out various coordinates to try to guess where your ships are. The object of the game is to sink each other's ships. The first player to do so wins the game.

**2.** Time how long it takes you to walk or ride the bus to school over a period of five days. Also time how long it takes you to get home over those five days. Find the average for both. Is there a difference? How many metres or kilometres would you estimate you walked or rode on the bus?

## Lesson 9
# Multiples and Division Rules

BUSINESS PLAN

PLAN:
**You will solve multiplication and division patterns.**

DESCRIPTION:
You have used multiplication and division to find rates, ratios, speeds, and distances that relate to your flyer route. In this lesson, you will extend your knowledge of multiplication and division using some of the numbers from your delivery route.

7, 14, 21, 28, 35

## Get Started

To find a multiple of a number, you count by that number. Here's an example:

In each set of flyers you deliver, there are seven flyers. If you deliver one set of flyers to each house, how many individual flyers would you deliver to five houses?

7,  14,  21,  28,  35

In this example, you would deliver 35 flyers in total. As we count by sevens, we name the multiples of 7. If we were to count by threes, we would name the multiples of 3.

1. Record in your notebook whether these numbers are odd or even.
   **a)** 1001    **b)** 234    **c)** 63
   **d)** 707    **e)** 556

2. List five multiples of each of these numbers:
   **a)** 3    **b)** 4    **c)** 6    **d)** 9    **e)** 12

3. List five multiples of your average walking speed that you figured out in Lesson 8.

4. Write all the numbers that will divide evenly into each of these numbers.
   **a)** 6    **b)** 7    **c)** 4    **d)** 12    **e)** 20

## Multiples and Division Rules

**1.** Write the next four multiples for each set of numbers.

    **a)** 2, 4, 6, ■, ■, ■, ■   **b)** 4, 8, ■, ■, ■, ■     **c)** 7, 14, ■, ■, ■, ■

    **d)** 10, 20, ■, ■, ■, ■   **e)** 15, 30, ■, ■, ■, ■   **f)** 11, 22, ■, ■, ■, ■

**2.** Write five multiples of the number 36.

**3.** The number 12 is the sixth multiple of which number?

**4.** Name three multiples of the number 60.

Read these division rules:

A number is divisible if, after the division, the remainder is 0.

A number is divisible by

- 2 if the last digit is even (0, 2, 4, 6, 8)
- 3 if the sum of the digits can be divided by 3
- 4 if the last two digits make a number that can be divided by 4
- 5 if the last digit is 0 or 5
- 6 if the number can be divided by both 2 and 3
- 9 if the sum of the digits can be divided by 9
- 10 if the last digit is 0

**5.** Use the division rules to decide if these numbers can be divided by 2 or 5. Identify numbers that can be divided by both 2 and 5.

    **a)** 82   **b)** 50   **c)** 55   **d)** 26   **e)** 17   **f)** 105

**6.** Use the division rules to decide whether these numbers can be divided by both 2 and 3.

    **a)** 36   **b)** 40   **c)** 49   **d)** 42   **e)** 60   **f)** 90

**7.** Review the division rules to find another number that would divide evenly into most of the numbers in question 2.

**8.** Find the numbers in questions 1 and 2 that prove the division rules for the numbers 4, 9, and 10.

## What Did You Learn?

Explain how using multiples can help you solve math problems. Give an everyday example where multiples would make problem solving easier.

## Practice

1. What are the common factors of 4, 8, and 12?

2. Create three questions about multiples for a classmate to solve.

3. If a number can be divided by 2 and 4, can it be divided by 8? Explain.

4. Create three questions based on the division rules for a classmate to solve.

### Extension

5. With a partner, make a game that involves questions about multiples and questions based on the division rules. Remember to create answer cards and include the rules and instructions for playing your game.

6. Use grid paper or a computer application to create a blank wordsearch chart. Instead of filling in your chart with words, use numbers. Give instructions with your chart that involve multiples and/or division rules. For example, you might write "Circle all the numbers that can be divided by both 2 and 3 in blue." Give your chart to a classmate to solve. Check his or her work.

### Booklink

**Alice in Pastaland** by Alexandra Wright (Charlesbridge: Watertown, MA, 2002). An imaginary trip through Pastaland provides Alice with opportunities to explore number concepts and basic arithmetic as she tries to help a white rabbit solve a math problem.

# Common Multiples and Factors

BUSINESS PLAN

PLAN:
You will learn about least common multiples and greatest common factors to help you solve problems.

DESCRIPTION:
You deliver flyers to 110 houses. Your friend delivers to 60 houses. A third friend delivers to 40 houses. In this unit, you will explore these and other numbers to find their least common multiples (LCM) and greatest common factors (GCF).

## Get Started

To find the least common multiple (LCM), write multiples of each number until a common one is found.

Here is an example of how to find the LCM of 10 and 25:

Multiples of 10: 10, 20, 30, 40, 50, 60 ...

Multiples of 25: 25, 50, 60, 70 ...

The LCM of 10 and 25 is 50.

### Vocabulary

**factor:** A whole number that is multiplied by another whole number to find a product
**greatest common factor (GCF):** The largest number that is a factor of two (or more) numbers; for example, 36 and 60 can both be divided evenly by 12. This is the GCF of both numbers.
**least common multiple (LCM):** The smallest number other than zero, that two or more numbers can divide into evenly; for example, the LCM of 2 and 3 is 6.

To find the greatest common factor (GCF), write the factors of each number and find the largest factor shared by both numbers.

Here is an example of how to find the GCF of 10 and 20:

10 = 1 x 10, 2 x 5

Factors of 10: 1, 2, 5, ⑩

20 = 1 x 20, 2 x 10, 4 x 5, 2 x 2 x 5

Factors of 20: 1, 2, 4, 5, ⑩, 20

The numbers 10 and 20 share these common factors: 1, 2, 5, and 10. The number 10 is the GCF of 10 and 20.

1. Find the LCMs and GCFs of these numbers. You may use a calculator to help you.

   a) 40 and 60

   b) 34 and 51

   c) 85 and 90

2. Find the LCMs and GCFs of these three numbers: 40, 60, 110.

## Build Your Understanding

### Find LCMs and GCFs

Use your knowledge of multiples to solve these word problems.

1. Margaret gets paid every 4 days. She goes to the bank to update her bank book every 14 days. How many days will pass before Margaret goes to the bank on the same day she gets her pay?

2. Li and Joey are taking part in a chess tournament. Li plays a match every 3 days. Joey plays a match every 5 days. On what day will they both play at the same time?

3. A pattern of geometric figures repeats every 10 figures. Another geometric pattern repeats every 8 figures. How many figures will be in each pattern before they repeat at the same time? Draw your answer.

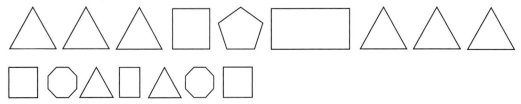

4. Ahanu and Phillip are making goodie bags for a party for the daycare children. They have 24 boxes of crayons, 36 markers, and 48 pads of paper. Show how they can divide the items to make the greatest number of goodie bags with nothing left over.

Find the LCMs and GCFs of these pairs of numbers:

**1.** 10 and 12     **2.** 11 and 22     **3.** 18 and 24     **4.** 20 and 24

**5.** Which of these numbers share the same LCM?

5, 8, 14, 16, 21?

## Extension

**6.** You can also use prime factors to find the LCM and GCF of sets of numbers. Work with a partner. Choose one or two sets of numbers, and follow these directions to find their LCM and GCF.

| To find the LCM of numbers: | To find the GCF of numbers: |
|---|---|
| **1.** Write the prime factors of each number. | **1.** Write the prime factors of each number. |
| **2.** If any factors of the first number also appear for the second number, cross them out for the second number. | **2.** Write the prime factors that the numbers have in common. |
| **3.** Multiply the numbers that remain to find the LCM. | **3.** Multiply the common prime factors. The product will be the GCF of the numbers. |
| **4.** Check your answers by listing multiples for each number. | |

### Show What You Know

**Review: Lessons 9 and 10, Division Rules and Common Multiples and Factors**

**1.** A dog breeder has nine dogs. Each dog has five puppies. How many puppies will the breeder have to find homes for?

**2.** Families won't adopt more than 3 puppies. If families adopt 3 puppies at a time, how many

families would be required to adopt all the puppies?

**3.** If $\frac{2}{3}$ of the puppies are females, and every family wants at least one female, how many families could adopt puppies?

# Chapter Review

**1.** Estimate an answer to each equation by rounding the numbers to the nearest 10 or 100.

**a)** 320 x 18 = ■

**b)** 510 + 265 + 317 + 792 = ■

**c)** 656 ÷ 41 = ■

**d)** 791 − 313 = ■

**e)** 37 + 52 + 98 + 19 = ■

**f)** Select and calculate one question from above. Explain why you underestimated or overestimated the answer.

**2.** There are 2 dogs, 6 cats, and 9 rabbits on Tony's farm. Write ratios for the following, then write each ratio as a fraction in lowest terms.

**a)** cats to dogs          **b)** rabbits to cats          **c)** dogs to cats

**3.** Write equivalent ratios for:

**a)** 4:5     **b)** $\frac{25}{5}$          **c)** 2:16     **d)** $\frac{120}{15}$

**e)** Draw a picture to show one of the ratios above.

**4.** The ratio of girls to boys at a school was 5:3. How many girls were at the dance if 84 boys were there?

**5.** Write the following percents as fractions and decimals.

**a)** 15%     **b)** 23%     **c)** 65%     **d)** 90%

**6.** Eric's brother, Jim, makes $6.00 an hour. He got a 4% pay increase. What is Jim's new unit rate? How much will Jim now make if he works 20 hours next month?

**7.** Arrange the following in order from least to greatest.

**a)** 1   4   9   5   7   0

**b)** 4.2   0.2   5.7   0.62   3.01

**c)** $\frac{15}{100}$   $\frac{4}{20}$   $\frac{25}{100}$   $\frac{1}{100}$   $\frac{5}{10}$

**d)** Plot one of the above sets of numbers on a number line.

**8.** Copy this coordinate grid into your notebook.

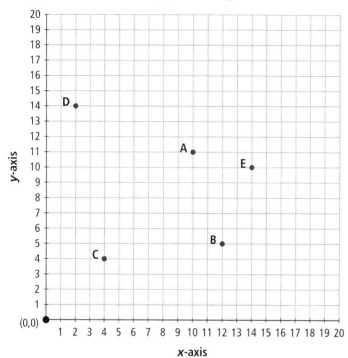

a) Record the ordered pair that gives the location of each of the five points (A, B, C, D, E) on the grid.

b) Which would be the shortest distance to travel: A to D or A to C? Explain your answer and draw your route. Remember, you must travel only along the grid lines.

**9.** Eric walked the following distances last week:

4.0 km,   5.0 km,   4.7 km,   5.4 km,   4.9 km,   6.1 km,   4.9 km

a) What was the average number of kilometres Eric walked?

b) If Eric walks at a rate of 5 km/h on average, about how many hours does Eric walk each week?

**10.** List the factors of 6 and 8. What is the greatest common factor (GCF) of 6 and 8?

**11.** List the first ten multiples of 5 and 20. What is the least common multiple (LCM) of 5 and 20?

# Chapter Wrap-Up

In this chapter, you have covered a number of topics, including

- ratios and rates
- percents and fractions
- number lines
- coordinates
- multiples and division rules
- least common multiples and greatest common factors

Now you can apply and develop your mathematical skills and knowledge.

Imagine that you have been asked to present a business plan to a group of Grade 5 students who will be starting their own flyer-delivery business next year. They will need to know about some of the things you have learned in this chapter.

1. On a coordinate grid, plot a point representing your house, plus 10 other points to represent the houses you have to deliver to on your route.

**2.** There are 80 flyers to be delivered. Calculate how many flyers will be delivered to each address, and express this as a ratio.

**3.** If each grid square is equal to half a kilometre, calculate how many kilometres your route is, starting and ending at your house. Draw a number line to represent the total distance, then plot points on the number line to show the approximate distances between the addresses.

**4.** If you walked at a speed of 4 km/h, calculate how long it would take you to complete your route.

**5.** Pick two distances on your number line and calculate the least and greatest multiples and factors for those two numbers.

**6.** Locate the halfway mark on your number line, and write as a fraction the number of houses on each side of that mark. What are these fractions as decimals? as percents?

**7.** Calculate how much money you would make per month, if you earned $0.05 for every flyer you delivered, and you did your route three times per week.

**8.** Create a tip sheet with at least five pieces of advice you would give the Grade 5 students when preparing their own flyer delivery business plan.

Organize your findings, then make a colourful front and back cover for your work. Share your presentation with the class.

# Chapter 8

# The Data and Probability of Business

In this chapter, you will continue exploring the way math is used in business.

In this chapter, you will

- develop and conduct surveys
- analyze population samples
- display data in various ways
- explore different kinds of probability
- determine the probability of events
- solve math problems by creating organized lists and by simplifying problems

At the end of this chapter, you will survey people in your school, then create a flyer based on those survey results.

Answer these questions to get you started:

1. Have you, or has anyone you know, ever taken part in a survey? What questions were asked? What were the answers used for?

2. Explain to a classmate what you know about problem solving.

**Lesson 1**

# Developing Surveys

*BUSINESS PLAN*

*PLAN:*
**You will develop a survey about one type of flyer and the products displayed in it.**

*DESCRIPTION:*
*Surveys are used to identify people's opinions about many different topics. The information gathered can be used in many ways.*

---

**Cat Food Survey**

1. What types of food do you feed your cat?

a) dry food only ☐

b) wet food only ☐

c) a combination of ☐
wet and dry food

---

## Get Started

During the next few lessons in this unit, you will be preparing and conducting a survey on one type of product. Map out the steps you think you will need to prepare and conduct the survey. For each task, list factors you will need to consider.

### Vocabulary

**survey:** A method of gathering information by asking questions and recording people's answers

## Build Your Understanding

### Develop a Survey

You Will Need
- various types of flyers
- magazines containing surveys

**1.** Your teacher will gather six types of flyers and group them into categories like these:
- drugstores
- food stores
- toy stores
- hardware stores
- electronics stores
- department stores

**2.** Choose a flyer on which you would like to base your survey.

**3.** Look at the products listed in the flyer and how various companies market their products.

**4.** Select one major topic that you would like to learn more about. Perhaps you want to find out what people look for in a flyer. Is it the range of products offered and discount prices, or is it a particular product's features, quality, and appearance?

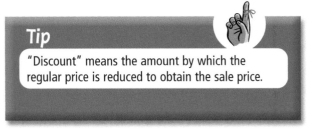

**Tip**

"Discount" means the amount by which the regular price is reduced to obtain the sale price.

**5.** Look at surveys in magazines. Note how the questions are phrased and the number of questions included in the surveys. Identify the audience for each survey.

**6.** Brainstorm a list of questions related to your topic.

**7.** Write your questions so that they can be answered in one or two words.

**8.** Organize your questions in a list.

**9.** Trade questions with a partner.

**10.** Check your partner's questions for
- fairness (is the survey trying to convince people about a particular product?)
- repetitiveness (are some of the questions too similar?)
- valuable information (are some questions unrelated to your survey topic? Are some of your questions vague or open-ended?)

1. How did you change your survey after your partner reviewed it?

2. What is the most important thing you hope to learn from your survey? What question will help learn this?

## Practice

1. Ask a partner to complete your survey. Time how long it takes him or her to answer your questions.

2. Complete your partner's survey as he or she times how long it takes you.

3. Provide each other with feedback on the surveys.

4. Make revisions to your work based on the feedback you have received.

You will use this survey in the next few lessons.

## Extension

5. Most flyers include prices. What other numbers can you find in the brochures you have read? Record the numbers, and classify them according to their purpose.

## Lesson 2

# Sample Populations

BUSINESS PLAN

PLAN:
You will select a sample population in order to complete a survey.

DESCRIPTION:
Surveys are given to a small group of people who represent a larger group.

## Get Started

Try dividing your class into groups according to various characteristics. Here are some ideas to get you started:

- students with pets
- students with siblings
- students who play a musical instrument

### Vocabulary

**population:** The whole group of objects or individuals considered for a survey, such as all Grade 6 students in Canada
**sample:** A part of a population selected to represent the population as a whole; for example, if the population is all Grade 6 students in Canada, a sample would be all Grade 6 students in your school

1. Discuss as a class how each group could represent a larger group of people. Would one group represent, for example, all people with siblings? Discuss why or why not.

2. Read these examples of survey topics. For each example, decide whether you would survey the entire population or just a sample.

   a) You want to find out how many students in your class have a math computer game.

   b) You want to discover if Grade 6 students in your school are more likely than Grade 5 students to own bicycles.

**c)** You want to find out how many people in your town or city have answered a survey.

**d)** This year, 300 people in your town or neighbourhood will buy an electronic device. You want to identify what device most people will buy.

## Build Your Understanding

### Identify a Sample Population

It is important to choose a sample that represents the population. Think about the population you want to interview about the flyer you have chosen. Make a list of characteristics that represent that population.

> Imagine you are surveying Grade 6 students about their favourite computer games. You would need to survey students who like to play computer games. (If someone doesn't like to play computer games, chances are that she or he won't have a favourite game.) Think of other characteristics that the people you survey (your target sample) should have.

Review your survey and identify the following factors:

- the number of people you will survey
- where you will conduct your survey
- characteristics of the people in your sample

## What Did You Learn?

1. Without rereading the text, define the terms "population" and "sample" in your own words.

2. Although most surveys use sample populations, some surveys use entire populations. What examples can you think of where an entire population would have to be surveyed?

1. Trade your answers from Build Your Understanding with a classmate. Review his or her answers. Analyze how well this sample represents the population. Suggest changes, if needed, and use your partner's suggestions to revise your own characteristics.

The following are examples of sample populations for surveys. For each example, decide how well the sample would represent the population.

2. Mahalia wants to find out if all students her age (12) like sports. She surveys 20 students about their interests in sports. All of these students play sports for their school.

3. Class 6B is holding a fundraising lunch. Lunch organizers must choose either pizza or hot dogs for the lunch. They survey all students in the class to find out what they would prefer to eat and what would encourage them to participate.

4. Mahalia wants to find out how many people in the school are going away for summer vacation. She asks every fourth person she meets in the school. She surveys 40 people.

**5.** The teacher wants to encourage people to return their library books on time. He conducts a survey of 50 users, some of whom always return their books late and some of whom always return their books on time. In his survey, he tries to identify strategies to help him achieve his goal—to have no overdue library books.

There are three types of samples: convenience, random, and systematic. Below are definitions of the three types. Write the matching definition beside each type of sample in your notebook.

**6.** Individuals in the population are located near the person administering the survey. This makes it easy and fast to conduct the survey.

**7.** Each individual in a population has an equal chance of being chosen. This type of sample is the most representative of a population.

**8.** One person is chosen randomly, and then other people are chosen according to a pattern (for example, every tenth person).

**9.** Identify which type of sample would provide the best results. Give reasons for your choice.

**10.** Review the sample populations in questions 2 to 5. For each population, identify the sample type (convenience, random, systematic).

**11.** Which type of sample did you use for your survey?

## Lesson 3

# Conducting a Survey

BUSINESS PLAN

PLAN:
You will conduct your survey with at least 15 people.

DESCRIPTION:
You have prepared your survey and identified your sample population, so now you're ready to go. It's time to survey!

## Get Started

This is your last chance to fine-tune your survey. You must make sure that none of your questions are written in such a way that they will influence the results of your survey. Such influence is called bias.

### Vocabulary

**bias:** An emphasis on characteristics not typical of the whole population

In a survey, a biased question might encourage people to answer in a certain way and therefore affect the results of the survey may not accurately reflect the characteristics of the sample. Here are some examples of biased questions:

**A.** Do you agree that volleyball is a better game than basketball?

**B.** Don't you think it would be best for all students to wear school uniforms?

**C.** Which Canadian province do you think is the most beautiful— British Columbia, Ontario, or Prince Edward Island?

Work with a partner.

**1.** Rewrite the questions above to remove bias. Discuss your revisions with your partner. What did you change to eliminate bias?

**2.** Go back to Lesson 2 and look at Practice question 2. Consider the likely results of Mahalia's survey. Would it give an accurate picture of sports interest among 12-year-olds? Why or why not? Discuss your ideas with a classmate.

## Build Your Understanding

### Conduct a Survey and Collect Data

Work with a partner.

**1.** Finalize your sample population (you should have at least 15 people in your sample). Reread the list of people you will survey to make sure your list is representative of the population. Make any necessary revisions before beginning your survey.

**2.** Conduct your survey according to your plan. Ensure that you record responses clearly and that you include the name of the person on the form with his or her responses.

**3.** Create a tally chart and record the results of your survey on it. This is an important step for Lesson 4, in which you will analyze and present your data.

## What Did You Learn?

Read the following survey questions:

**A.** Do you agree that volleyball is a better game than basketball?

**B.** Don't you think it would be best for all students to wear school uniforms?

**1.** Mahalia wants to know the favourite ball game among 12-year-old sports participants. Would question A be useful to her? Would it be useful if she wanted to know which of the two ball games is preferred among her chosen population? Who should she include in her sample to represent the population she wants to survey?

**2.** Eric heard a lot about school uniforms and decided to find out how many students wanted the option of wearing one. Question B is the first draft of his survey question. How might he reword it to eliminate bias?

**3.** Biased questions will negatively affect the results of a survey. Explain why.

## Practice

**1.** Review surveys in magazines. Compare the data in these surveys to the data from your survey. Make note of graphs and other devices people use to share the results of their survey.

**2.** In some surveys, you may notice these terms: mean, median, and range. In your notebook, match these definitions to the terms.

   **a)** The middle value in a list of numbers arranged in order by size.

   **b)** The average of a set of amounts.

   **c)** The number that represents the difference between the greatest and least values in a group.

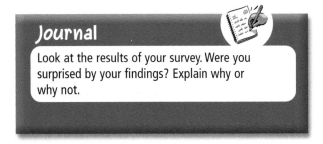

### Journal
Look at the results of your survey. Were you surprised by your findings? Explain why or why not.

### Show What You Know

**Review: Lessons 1 to 3, Surveys**

**1.** Using the Internet, magazines, and newspapers do some research to find out how surveys are used in real life. Make note of any examples of bias that you find in the surveys.

**2.** Make a graph. Develop a question for a classmate to answer.

## Lesson 4

# Presenting Data

*BUSINESS PLAN*

*PLAN:*
**You will learn about a variety of graphs and tables.**

*DESCRIPTION:*
In the last few lessons, you collected and organized data for your survey. Now it's time to graph it. Graphs are a visual way of representing your findings. You've probably seen graphs and tables, like the ones that follow, in magazines or textbooks. Graphs and tables allow information to be given in a small area, without long text explanations.

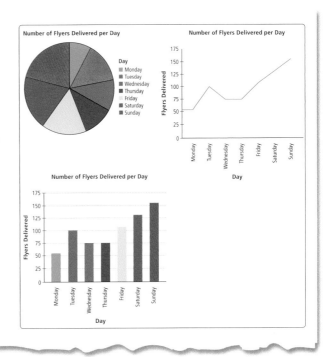

## Get Started

In one school, 68% of all students have grandparents who live in the same town or city, 22% have grandparents who live in another town or city, and 10% have grandparents who live in another province or country. This data has been displayed on a circle graph and a bar graph.

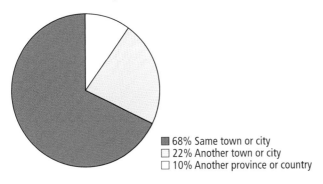

Where Students' Grandparents Live

■ 68% Same town or city
□ 22% Another town or city
□ 10% Another province or country

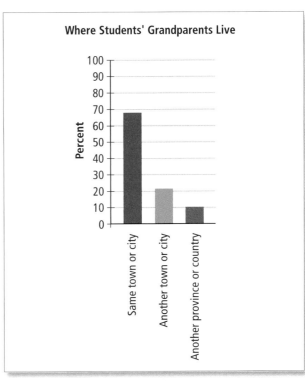

Where Students' Grandparents Live

Lesson 4: Presenting Data

357

What does each graph tell you? Which graph is easiest for you to understand? Which of these graphs do you find most effective? Why?

Now look at the survey results you tallied in Lesson 3. To help you decide on the graph type(s) to represent your data, you might want to turn your tallies into tables.

One type of table is a frequency table, which gives you the total for each category or group. Here is an example to get you started:

One of Anna's questions in her survey asked people how they would rate the flavour of one brand of ice cream. This table shows their responses.

| Frequency Table | |
| --- | --- |
| Rating the Flavour of One Brand of Ice Cream | |
| Excellent | 6 |
| Very good | 4 |
| Average | 3 |
| Below average | 2 |
| Poor | 0 |

Another type of table that is useful for organizing information is an interval table. In an interval table, you group numbers to make data easier to understand and evaluate. For example, Ahanu wanted to know how many times per month people bought products at one store. Here are the responses: 0, 3, 4, 6, 7, 8, 10, 12, 13, 15, 18, 18, 20, 22, 25. Using this information, he prepared the following interval table:

| Interval Table | | | | | |
| --- | --- | --- | --- | --- | --- |
| People Buying Products at One Store | | | | | |
| Times per Month | 0–5 | 6–10 | 11–15 | 16–20 | 21–25 |
| | /// | //// | /// | /// | // |

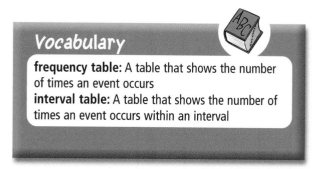

**Vocabulary**

**frequency table:** A table that shows the number of times an event occurs
**interval table:** A table that shows the number of times an event occurs within an interval

## Build Your Understanding

### Plan Your Data Presentation

Look at the chart below. It contains the most common types of graphs seen in newspapers, magazines, and journals. Read the description and look at the example for each type of graph.

| Type | Description | Example |
|---|---|---|
| bar graph | • best suited to show one set of data<br>• bars can be horizontal or vertical but are usually vertical<br>• bars are separated, not connected | 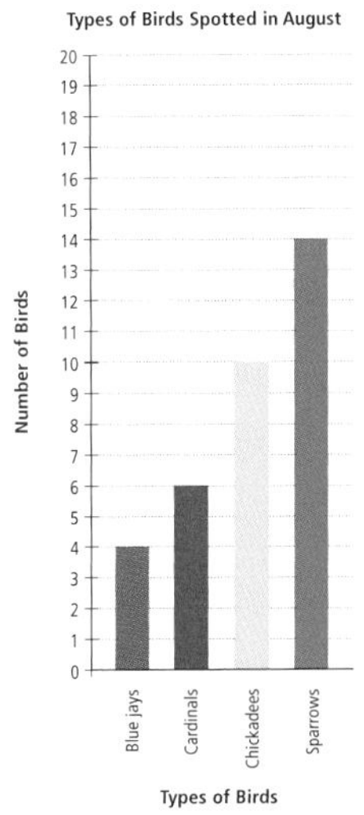 |
| double-bar graph | • used to show two sets of data on the same topic<br>• double bars should be separated, not connected | 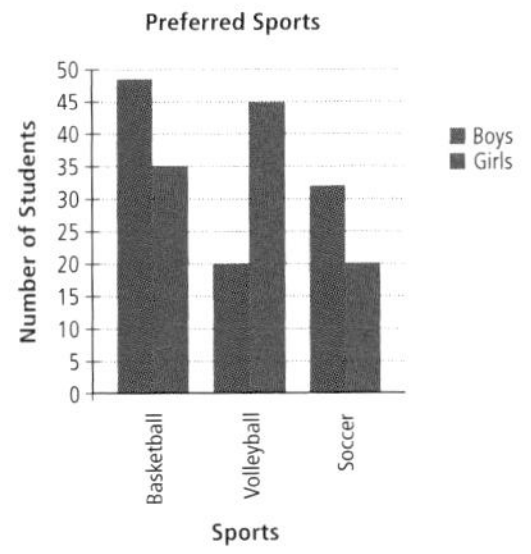 |
| line graph | • used to show changes over a period of time | 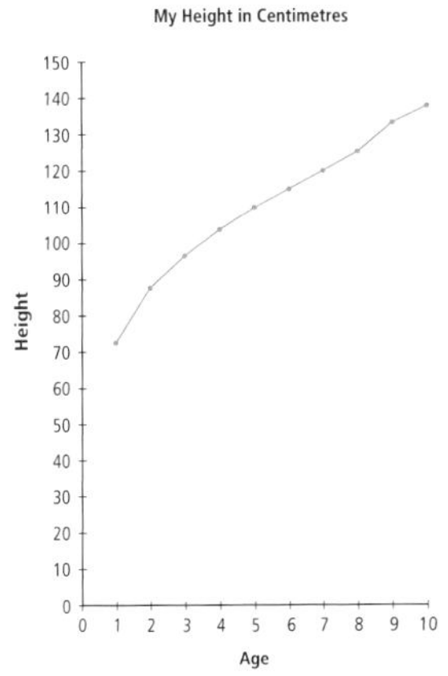 |

| Type | Description | Example |
|------|-------------|---------|
| double-line graph | • used to display two sets of data over a period of time | 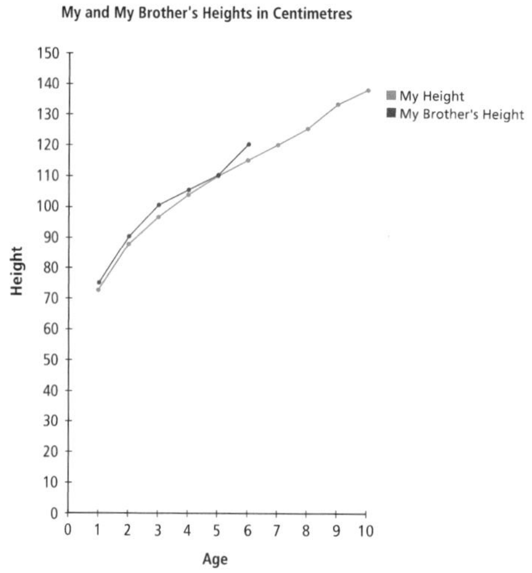 **My and My Brother's Heights in Centimetres**<br><br>My Height / My Brother's Height<br>Height vs Age |
| circle or pie graph | • best used when comparing parts to a whole or to other parts | 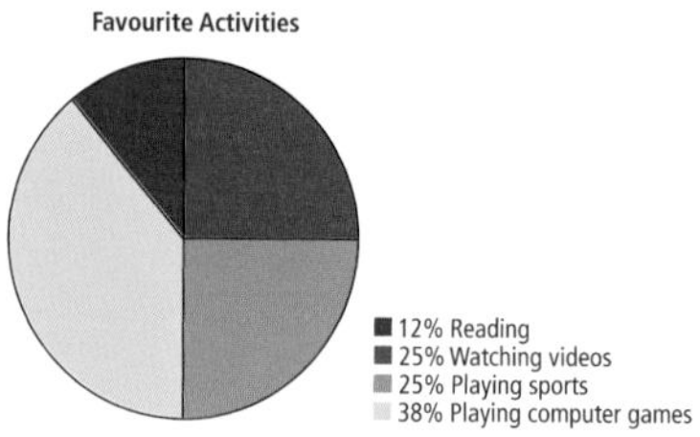 **Favourite Activities**<br><br>■ 12% Reading<br>■ 25% Watching videos<br>■ 25% Playing sports<br>■ 38% Playing computer games |
| histogram | • best suited to show the frequency or number of times data occur within intervals<br>• bars are connected | 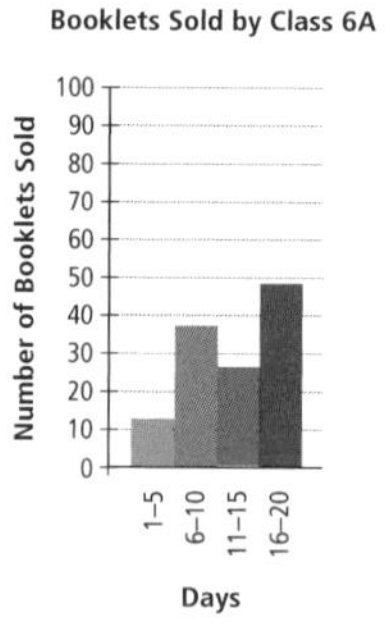 **Booklets Sold by Class 6A**<br><br>Number of Booklets Sold vs Days (1–5, 6–10, 11–15, 16–20) |
| stem-and-leaf plot | • best used when it is important to see each number in a set of data<br>• stem numbers are written vertically; leaf numbers are written horizontally<br>• usually, the tens go in the stem column; the ones go in the leaves column | 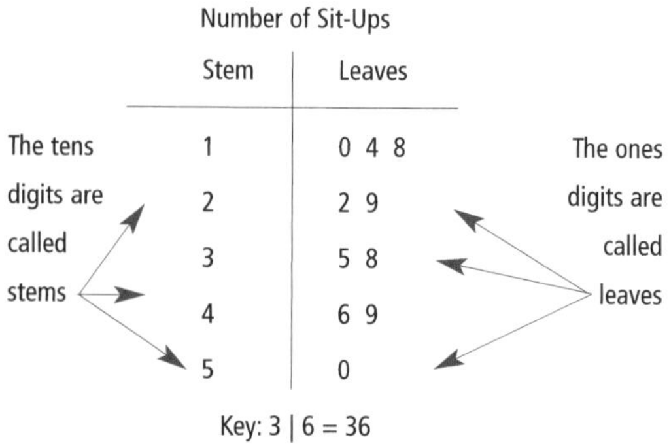 **Number of Sit-Ups**<br><br>The tens digits are called stems<br><br>| Stem | Leaves |<br>\|------\|--------\|<br>\| 1 \| 0 4 8 \|<br>\| 2 \| 2 9 \|<br>\| 3 \| 5 8 \|<br>\| 4 \| 6 9 \|<br>\| 5 \| 0 \|<br><br>The ones digits are called leaves<br><br>Key: 3 \| 6 = 36 |

Experiment with your data using a computer, or draw quick sketches. Try different forms of graphs to find out which of the above graph types would represent your data most effectively.

## What Did You Learn?

1. Name the type of graph you see used most often and where you see it.

2. What type of graph would you use to show the temperature every day for a month? Why?

3. What type of graph would you use to show the number of students in your class who walk, are driven, or ride bikes to school? Why?

## Practice

1. With a partner, discuss your findings and any difficulties you had summarizing your data.

2. Describe how the size of intervals can affect the appearance of data.

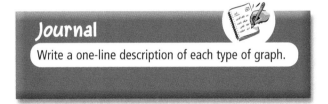

**Journal**

Write a one-line description of each type of graph.

## Extension

3. Combine your results with the survey information from four or five classmates. As a group, write, edit, and illustrate a Tips Booklet. Include information that you think someone would need to prepare and carry out a survey, and to present findings from this survey to an audience.

4. Add to your Tips Booklet any helpful information you learned about representing information using various types of graphs. Include what you know about how the size of intervals can affect the appearance of the results.

## Lesson 5

# Designing a Data Presentation

### BUSINESS PLAN

**PLAN:**
You will apply all that you have learned in the last few lessons to create your own data presentation.

**DESCRIPTION:**
Choosing the most appropriate type of graph to represent your work is very important. Some information is better suited to certain types of graphs.

## Get Started

Look at the examples below. For each example, identify the type of graph or table that would best represent the data.

**Tip**

For a few examples, you could use more than one type of graph.

1. video prices over a five-year period

2. scores out of 100: 19, 25, 26, 34, 45, 54, 63, 65, 68, 72, 78, 80, 88, 90

3. number of times per year that male and female students
   a) go to the movies
   b) rent a video
   c) see a live performance (play, concert)

4. number of items in a store that range in price from $0.01 to $0.20, $0.21 to $0.40, $0.41 to $0.60, $0.61 to $0.80, and $0.81 to $1.00

5. number of days each month Ruth attended swimming lessons in a six-month period

6. the number of cities with populations between 50 000 and 180 000

## Build Your Understanding

### Create a Graph to Display Your Data

Now that you have completed all the background work, it's time to prepare your data presentation. You may be able to group responses to some questions together, while other questions will require their own graph. As you work, keep these rules in mind:

**Technology**

Use computer applications, if possible, to prepare your work.

1. Match your data to the most appropriate type of graph.

2. Label and title your graphs clearly.

3. Enter your data on the graphs so that it is easy to understand.

When you are finished, assess your work. How well does it tell your audience the findings of your survey? Ask for comments and suggestions from three classmates. Consider their comments when preparing your final presentation. In addition to the graphs, provide information for your audience about your survey, including characteristics of your sample population, where and when you conducted your survey, and the type of survey you performed.

## What Did You Learn?

1. Describe your experience of grouping responses. What factors made the experience easy? What made it difficult?

2. Think about the comments of your classmates. How were they able to help you improve your work?

1. With your teacher's help, create a classroom data presentation display.

2. Provide additional information with your graph(s) so that readers understand the context of your survey.

3. Look at some of your classmates' displays. From the material provided, what information can you learn from their surveys? Identify factors that help you understand a visual presentation of data.

## Extension

4. Over a one-week period, scan a newspaper each day. Keep a daily record of the graphs used in the newspaper. For each graph, summarize the information it presents. At the end of the week, review your notes. Identify the graph type used most often, and give reasons why you think this is the case.

## Show What You Know

### Review: Lessons 4 and 5, Presenting Data

Enter the data from the Extension or other data into a computer program. Make different types of graphs. Which are the easiest to make on the computer? Which are the most difficult to make?

## Lesson 6

# Exploring Probability

*BUSINESS PLAN*

*PLAN:*
You will learn about the probability of some events occurring.

*DESCRIPTION:*
In previous lessons in this chapter, you worked with different types of flyers. How many students chose to work with the same type of flyer? There are two ways you can find out: by surveying your classmates and tallying the numbers, or by figuring out the probability that more than one student made the same decision.

## Get Started

When you work with probability, you also work with fractions, decimals, and ratios.

**1.** Convert the following fractions to decimals:

a) $\frac{34}{100}$    b) $\frac{3}{5}$    c) $\frac{14}{20}$    d) $\frac{100}{100}$    e) $\frac{114}{100}$

**2.** Convert the following decimals to fractions:

a) 0.61    b) 0.02    c) 1.1    d) 0.99    e) 1.5

**3.** Write each fraction in its lowest terms:

a) $\frac{75}{100}$    b) $\frac{20}{50}$    c) $\frac{50}{60}$    d) $\frac{102}{100}$    e) $\frac{32}{48}$

### Vocabulary

**probability:** The chance or likelihood that an event will happen

## Express Probability as Fractions and Percents

In this example, if one colour is chosen, red and white have an equal chance of being chosen. They have equal outcomes. The two colours make up the sample space of all possible outcomes in that set of colours. The probability of selecting one of the two possible colours is $\frac{50}{100}$ or $\frac{1}{2}$ or 50%.

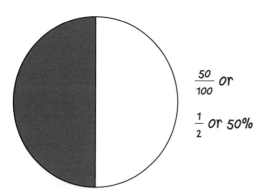

$\frac{50}{100}$ or

$\frac{1}{2}$ or 50%

> **Vocabulary**
>
> **outcome:** A possible result of a probability experiment
> **sample space:** The set of all possible outcomes

For the following given outcomes, write the probability as a fraction and as a percent.

**1.** The colour blue will be chosen.

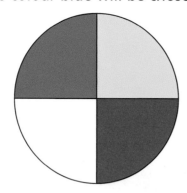

**2.** The colour yellow will be chosen.

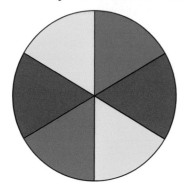

**3.** The colour green will be chosen.

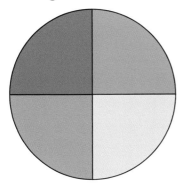

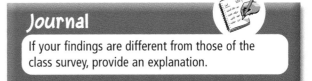

**Journal**

If your findings are different from those of the class survey, provide an explanation.

**4.** The colour red will be chosen.

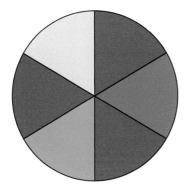

**5.** What are the chances that the person sitting beside you worked with the same type of flyer that you worked with? Use your knowledge of outcomes to figure out what the probability would be. Show your answer as a fraction and as a percent.

**6.** Once everyone has figured out the probability, conduct a class poll to find out the percentage of people in the class who used each type of flyer. Compare your findings with those of the class survey.

## What Did You Learn?

**1.** Without rereading your text, define these terms: probability, outcome, and sample space.

**2.** Name three games that are based on probability rather than skill.

Li is taking the bus to see her grandparents. There are 3 local buses and 1 direct bus. Usually, equal numbers of people ride on each bus.

**1.** What is the probability that Li will get a seat on the direct bus?

There are 10 chocolate bars in a bag. 3 bars are caramel-flavoured, which Mahalia likes, 2 are fudge-flavoured, which she dislikes, and 5 are orange-flavoured, which she likes. Mahalia put her hand into the bag to take out a chocolate bar without looking inside the bag.

**2.** What are her chances of pulling out a chocolate bar that she dislikes?

**3.** Write another probability question about the chocolate bars in the bag.

## Extension

**4.** Write rules to help someone who has difficulty converting percentages to fractions, fractions to percentages, and decimals to percentages. Provide examples to help explain your rules.

# Theoretical Probability

BUSINESS PLAN

PLAN:
**You will learn about theoretical probability by using number cubes.**

DESCRIPTION:
Think about a time when you wanted the weather to be clear—no clouds and no rain. When you were calculating the chance of clear weather, you were exploring theoretical probability— comparing the number of favourable outcomes to the number of possible outcomes. In this case, the favourable outcomes were no clouds and no rain; the other outcomes were rain and clouds.

## Get Started

Probability is the chance that an event will occur. The probability of an event can be written as a fraction, a decimal, a percent, or a ratio. Use the following equation to identify the probability of the spinner landing in the red section:

$$P\text{ (red)} = \frac{1 \text{ favourable outcome}}{2 \text{ possible, equally likely, outcomes}} = \frac{1}{2} = 50\%$$

The probability that the spinner would land in the section is $\frac{1}{2}$ or 50%.

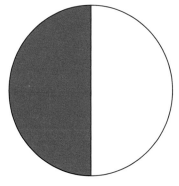

### Vocabulary

**theoretical probability:** The number of favourable outcomes compared to the number of possible outcomes when all outcomes are equally likely

### Find the Theoretical Probability of Game Events

You Will Need
• number cube

Work with a partner.

**1.** Using a number cube, calculate the probability of rolling a number that is

   **a)** less than 6 but greater than 3

   **b)** more than 1

   **c)** even

   **d)** the same as the last number rolled

> For each probability, write an equation that includes a fraction, a decimal, and a percent. Here is one example that shows a probability of 1 in 3:
>
> $P (1 \text{ or } 2) = \frac{2}{6} = \frac{1}{3}, 0.33,$ or 33%

To find the probability of an event *not* occurring, you must calculate:

$1 - P$

> Using the example of the red and white spinner in Get Started, if you wanted to find the probability of the spinner *not* landing on the red section, you would write the following:
>
> $P (\text{not red}) = 1 - P (\text{red}) = 1 - \frac{1}{2} = \frac{1}{2}$
>
> $P (\text{not red}) = \frac{1}{2}, 0.50,$ or 50%
>
> When rolling a number cube, the probability of rolling a 3 is $\frac{1}{6}$. The probability of rolling each of the other numbers is also $\frac{1}{6}$; so the probability of *not* rolling a 3—of rolling a 1, 2, 4, 5, or 6—is $\frac{5}{6}$.
>
> What do you notice about the probability of rolling 3 and not rolling 3? $\frac{1}{6} + \frac{5}{6} = 1$. So to find the probability of *not* rolling 3, we could find the probability of rolling 3 and subtract it from 1.

**2.** Find the probability of the events in question 1 *not* occurring.

## What Did You Learn?

Could you use theoretical probability to find the probability of it raining tomorrow? Why or why not?

## Practice

1. Tony needs to roll a number less than 3 on a number cube. What is the probability that he will do this?

2. In a box, there are 19 small paper clips, 20 large paper clips, and 11 super clips. If Tony were to reach in the box and pull out the first paper clip he touched, what are the chances that he would take out

   a) a super clip?

   b) a large clip?

   c) a small clip?

3. Show your answers to question 2 as fractions, decimals, and percents.

### Booklink

**Conned Again, Watson** by Colin Bruce (Perseus: Cambridge, MA, 2001). Twelve stories created in the likeness of the original Sherlock Holmes adventures take the reader through logic, probability, game theory, and decision theory as they apply to everyday scenarios.

4. For question 2, find the probability of *not* taking out

   a) a super clip

   b) a large clip

   c) a small clip

5. Show your answers to question 4 as fractions, decimals, and percents.

## Lesson 8

# Experimental Probability

BUSINESS PLAN

PLAN:
You will find the experimental probability of an event and compare experimental probability with theoretical probability.

DESCRIPTION:
Experimental probability is another way of predicting probability. When you test to see how many times you can throw a piece of paper into a trash can from a distance of 2 m, you are working out the experimental probability of an event.

## Get Started

You Will Need
• paperclip
• pencil
• coloured pencils

Make a spinner containing five differently coloured sections. Choose one colour, and predict the likelihood of spinning that colour. Spin at least 10 times and record the number of times you landed on your colour. Write your result as a fraction out of the total number of spins. How does your data compare to your prediction? Compare your results with a classmate.

When you perform an experiment like this one (repeating the same action many times), you find the experimental probability of that event.

### Vocabulary

**experimental probability:** The chance of an event happening based on the results of an experiment

## Find the Experimental Probability of Events

Working with a partner, decide on a game that you can use to find the experimental probability of an event. Here are some ideas to get you started:

• drawing one letter of a word from a container

• spinning the same number(s) or colour(s) in a row

• flipping a coin and obtaining heads

To help you keep track of your observations, prepare a frequency table. Label the columns with all possible outcomes. Here is an example of a frequency table that could be used with the spinner on the right.

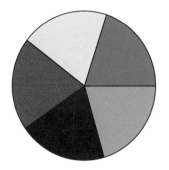

| Colour | Yellow | Blue | Green | Red | Black |
|---|---|---|---|---|---|
| Number of Spins | 5 | 6 | 5 | 6 | 3 |

This experiment was to find the probability of spinning yellow. The spinner was spun 25 times.

$P$ (yellow) $= \dfrac{5}{25} = \dfrac{1}{5} = 0.20 = 20\%$

Remember to write probability equations and to include fractions, decimals, and percentages in your equations.

Create a bar graph showing the probability of the outcomes in your experiment.

## What Did You Learn?

**1.** Flipping coins can be an example of both theoretical probability and experimental probability. Explain why.

**2.** Name two ways you use experimental probability in your own life.

1. Review the results of your activity. Discuss with a classmate why your results may be different from the probability that you expected.

2. Ajay makes a prediction of the time he will take to run 100 m over a period of 10 days after timing a friend run the same distance. How could Ajay make a more accurate prediction?

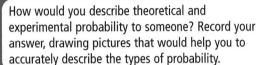

**Journal**

How would you describe theoretical and experimental probability to someone? Record your answer, drawing pictures that would help you to accurately describe the types of probability.

## Extension

3. Prepare at least two experimental probability problems for a partner. Provide your partner with a fraction, decimal, or percent equation, but do not identify the event. Use illustrations and objects to help you create your problems. Your partner must identify what the event might have been.

### Show What You Know

Review: Lessons 6 to 9, Probability

1. What is the probability of rolling doubles using two number cubes? Make a chart, then roll the number cubes 20 times. Record the results of your rolls.

2. What number comes up most frequently?

3. What number comes up the least frequently?

4. What patterns do you see? Record your observations in your journal.

## Lesson 9

# Creating Organized Lists

*BUSINESS PLAN*

*PLAN:*

**You will make organized lists to help you solve problems.**

*DESCRIPTION:*

*Solving any kind of math problem can be easier if you use tools to help you. In the next few lessons, you will learn about (and practise using) several math problem-solving tools.*

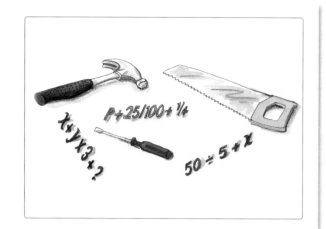

## Get Started

Li spun her spinner 5 times. Each time she spun, she recorded the number she landed on. Her total score was 15.

How many ways could Li have reached a total of 15 after 5 spins?

Record your answer in your notebook.

## Use Organized Lists to Solve Problems

**1.** Ajay wanted to buy fresh fruit for his friend and himself. He
had $6.90 in his wallet. Here are the costs of the fruits:

banana $0.40          orange $0.25

pineapple $5.25          apple $0.35

Ajay spent all of his money. Make an organized list to find out
what he may have bought.

**Journal**

Show how many different ways the letters A, B,
and C can be combined differently using all three
letters. Why do you think some provinces added a
fourth letter to car licence plates?

**2.** Mahalia was planning her sports schedule.
Her soccer games were scheduled for
Monday afternoon, Wednesday morning,
Thursday afternoon, and Friday morning.
Her swimming classes were offered on
Tuesday afternoon and Friday morning.
Mahalia was already enrolled in a crafts
class for Thursday afternoon. On what
days could she attend her soccer games?
On what days could she attend her
swimming classes?

**3.** Karen is a hockey referee. During a
two-month period, she refereed at games
where 120 goals were scored. In 2 of
these games, no goals were scored. The
average number of goals for each team
was 4 per game. How many games did
Karen referee?

## What Did You Learn?

**1.** Organized lists are one way we can use pencil and paper to help us organize our thoughts and record information. What other ways do you use written records to help you find answers and solve mathematical problems?

**2.** The problems in Build Your Understanding are word problems. Do you find them easier or more difficult to solve than math problems that are written in the form of equations? Why?

## Practice

### Math Problems to Solve

We often use math without even thinking about it. Take a few minutes to think about problem-solving strategies you use to help you solve everyday problems.

**1.** What type of problem would be difficult to solve using organized lists? Record reasons for your thinking, and then share your work with a classmate.

**2.** Mahad, Sheryl, Ezra, and Miriam went to Li's house to watch a movie. Each friend arrived within minutes of one another. Make an organized list to find out the order in which Li's friends may have arrived. How many different orders did you find?

**3.** Write a math problem for a classmate to solve. Your problem must be best solved using an organized list.

### Booklink

**Anno's Hat Tricks** by Anno Mitsumasa (Philomel Books: New York, NY, 1985). Three children, Tom, Hannah, and Shadowchild, who represents the reader, are made to guess, using the concept of binary logic, the color of the hats on their heads. This is an introduction to logical thinking and mathematical problem solving.

# Simplifying Problems

BUSINESS PLAN

PLAN:
**You will solve problems by making them simpler.**

DESCRIPTION:
In the last lesson, you learned about organized lists. Another good problem-solving strategy is making a problem simpler before you try to solve it. There are several ways you can do this. You can sort out important information from information that gets in the way of problem solving. You can also break the problem down to make it simpler. Once you solve the simpler problem, you are on your way to solving the more difficult one!

## Get Started

Li's younger sister has a problem. She wants to make a beaded necklace of three colours. She has 30 red beads, 20 blue beads, and 10 yellow beads. The string is yellow. Li's sister doesn't want to have two beads of the same colour together, and the necklace has to be loose. Li listens to her sister and notes the important information:

• three colours of beads—30 red, 20 blue, 10 yellow

• two beads of the same colour can't touch

Li uses coloured game discs to make a model of her sister's necklace. She knows she has to use twice the number of blue beads as yellow beads and three times the number of red beads as yellow beads.

She decides to start by using only 1 yellow disc, 2 blue, and 3 red. She experiments with different arrangements.

The final model that Li made is on the next page. Is there another way to make a similar necklace?

Li ignored information that wasn't important (the colour of the string, the looseness of the necklace), and she made the problem simpler by using a small number of discs to establish her pattern.

## Build Your Understanding

### Solve a Simpler Problem

Solve the following problems by making organized lists, identifying important information, and making it simple.

1. Phillip wanted to increase the distance he ran over a period of a year. He wanted to run half the distance on pavement in the city and the other half on gravel in the country. Phillip decided to run 1 km more every 2 weeks for 2 months, then 2 km more every 2 weeks for the next 2 months. After that, Phillip ran the same number of kilometres each week for the rest of the year. How many more kilometres did Phillip run during that year?

2. Mahalia and her sister wanted to buy a CD player. The model they wanted to buy was $149.99. They could put $50.00 down as a deposit and pay the rest in 4 equal installments. How much did Mahalia and her sister owe on the CD player after paying the deposit?

3. Claire's father is worried. He is taking a train to Windsor, where he has to catch another train. The train must make 8 more stops before it arrives at the Windsor station. Each stop is 7 min long. The actual travelling time is 1.5 h. He has 3 h to catch his connection in Windsor. Will he arrive in time? Show your work.

## What Did You Learn?

**1.** This lesson focuses on making it simple, while the previous lesson focused on organized lists. In this lesson, another problem-solving strategy was described. Name the strategy.

**2.** How did you decide if a piece of information was important? Share with a classmate, and compare how each of you made your decisions. Give reasons that might explain differences.

## Practice

A cafeteria manager wants to open one hour earlier every day to catch more of the rush-hour traffic. He estimates that he will make $300.00 in sales in that hour. Most of the money will come from selling coffee. He must pay 4 staff members an additional $12.00 each for that extra hour.

**1.** After paying the staff, how much more will he make each week? each month? each year?

### Extension

Look at advertisements in your local newspaper. Examine at least five ads. For each ad, compare the number of important facts to unimportant facts. Discuss your findings with a classmate.

**2.** Write a report about what you found out and what you think of your results. Use your data to prove your point.

### Journal

Make a list. On one side of your list, include facts that you need to solve the problem; on the other side, include facts that you don't need.

| Important Facts | Unimportant Facts |
|---|---|
|  |  |

### Booklink

**More Sideways Arithmetic from Wayside School** by Louis Sachar (Scholastic: New York, NY, 1994). Fifty-eight seemingly random problems arranged into 14 sections are presented in silly scenarios, but when the reader takes a closer look, they make sense, and problem solving can take place.

# Chapter Review

1. You want to survey students in your school to gather data on student preferences in after-school activities.

   a) Write 3 questions you would use to gather your data.

   b) Explain whether you would survey the entire population or a sample population. Describe in detail the population you would select.

   c) Which type of sample (convenience, random, or systematic) did you use?

   d) Describe how you would record the data collected.

2. Consider this this survey question:

   Do you agree that chess is a better activity than badminton?

   Is this a biased or unbiased question? Explain your answer.

3. Look at this tally chart.

| Days of Week | Hours of TV Watched |
|---|---|
| Sunday | //// |
| Monday | // |
| Tuesday | / |
| Wednesday | // |
| Thursday | |
| Friday | /// |
| Saturday | ##/ |

   Display this information in a graph. Name the type of graph used. Explain why you chose the graph you did.

4. Li wanted to know how many times per year students went to a movie theatre. Here are the responses:

   0, 3, 5, 7, 8, 9, 12, 13, 14, 15, 15, 17, 18, 20, 20, 21, 21, 23, 25

   Using this data, prepare an interval table.

**5.** Convert the following fractions to decimals:

**a)** $\dfrac{102}{100}$    **b)** $\dfrac{20}{50}$    **c)** $\dfrac{4}{5}$    **d)** $\dfrac{7}{20}$    **e)** $\dfrac{120}{100}$

**6.** Convert the following decimals to fractions. Make sure they are written in lowest terms. Then write each fraction as a percent.

**a)** 0.75    **b)** 1.25    **c)** 0.38    **d)** 0.06    **e)** 0.15

**7.** Create a blank spinner like the one shown.

**a)** Colour the spinner so that blue is more likely to occur than yellow, and yellow is more likely to occur than green.

**b)** Write the approximate probability for each outcome as a fraction and as a percent. You may use a calculator to help you.

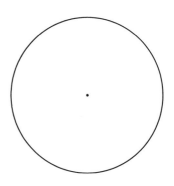

**8.** Look at this frequency table.

| Colour | red | blue | black | yellow | orange | white |
|---|---|---|---|---|---|---|
| Number of Spins | 9 | 9 | 8 | 7 | 10 | 7 |

**a)** Write three probability equations for the outcomes listed above. Include fractions, decimals, and percentages in your equations.

**b)** Is this an example of theoretical probability or experimental probability? Explain your answer.

**9.** Imagine you have some red paint, some black paint, and some yellow paint. You want to paint as many different cubes as possible. You can only use one colour per face. Three faces will always be without colour. How many different cubes can you paint? Create a net of the cube to help you. Show all of your work.

**10.** Form a group of 6 classmates. How many different

    **a)** pairs can you make up?

    **b)** groups of 3 can you make up?

**11.** Solve the following questions:

    **a)**
$$\begin{array}{r} 691 \\ 302 \\ 615 \\ +\ 373 \\ \hline \end{array}$$

    **b)**
$$\begin{array}{r} 7692 \\ -\ 1854 \\ \hline \end{array}$$

    **c)**
$$\begin{array}{r} 309 \\ 234 \\ 718 \\ +\ 865 \\ \hline \end{array}$$

    **d)**
$$\begin{array}{r} 325 \\ \times\ 28 \\ \hline \end{array}$$

    **e)**
$$\begin{array}{r} 73250 \\ -\ 1564 \\ \hline \end{array}$$

    **f)**
$$\begin{array}{r} 761 \\ \times\ 37 \\ \hline \end{array}$$

    **g)** $41\overline{)2214}$

    **h)** $15\overline{)4920}$

# Chapter Wrap-Up

In this chapter, you have covered a number of topics, including

- surveys
- population samples
- data display
- probability
- organized lists
- simplifying problems

Now you will have an opportunity to apply what you have learned.

You decide you want to create your own flyers about school events. Before you create them, you want to survey people in your school to see what sort of flyer they would like to see.

1. Survey a sample population in your school. You could include teachers, staff, or other students. Create a list of at least 10 questions to ask them about what kind of flyer they'd like to see.

2. Put the information you gather into an organized list. Create categories in which to put the information.

3. Based on your survey, decide what kind of flyer would best suit the people in your school. What is the probability that everyone you surveyed will be satisfied with your choice?

4. Create an organized list of products or information your flyer will contain. Do a rough layout of your flyer and share it with a classmate. Ask him or her to suggest changes for improvement. Incorporate the changes you agree with.

5. Once your flyer is complete, write the problems you encountered in its creation. Simplify the problems, then write solutions that might make the problems easier to solve for the next time you create a flyer.

# Chapter 9

# The Shapes and Patterns of Business

In this chapter, you will explore area, volume, mass, patterns, and geometry to help you decide how to spend your money.

In this chapter, you will

- study the relationships between measurements
- find areas of polygons
- find volumes of prisms
- find masses of various objects
- find the value of missing terms
- plot pairs of coordinates on a grid
- predict and represent transformations
- create tessellations

You will wrap up your work in the chapter by designing CD covers. You will also make and price containers for shipping the covers to customers.

Answer these questions to get you started:

1. What have you learned about measurement this year? How does what you have learned help you in everyday life?

2. How does what you know about geometry help you?

**Lesson 1**

# Volumes of Rectangular Prisms

BUSINESS PLAN

PLAN:
**You will estimate and then calculate the volumes of rectangular prisms.**

DESCRIPTION:
With some of the money you earned, you want to fill a box with chocolate as a gift for a friend. How can you figure out how many chocolates the box will hold?

## Get Started

To get ready for this lesson, estimate and then calculate the answers to these multiplication questions.

**1.** 4 x 3 x 2 = ■   **2.** 10 x 1 x 4 = ■

**3.** 8 x 6 x 2 = ■   **4.** 16 x 2 x 3 = ■

**5.** 20 x 2 x 3 = ■   **6.** 15 x 5 x 2 = ■

**7.** 90 x 5 x 2 = ■   **8.** 0.5 x 0.5 x 2 = ■

**9.** 400 x 3 x 6 = ■   **10.** 2 x 3 x 4 = ■

### Vocabulary

**rectangular prism:** A three-dimensional figure in which all six faces are rectangles; a square prism is a rectangular prism with two square faces

## Calculate the Volumes of Rectangular Prisms

**1.** The formula for volume is
$V = l \times w \times h$. Use this formula to find
the volume of each of the following
rectangular prisms. Make an estimate
first, and then calculate your answer.
Use your estimate to decide if your
calculations are accurate.

**Vocabulary**

**volume:** The amount of space that an object
occupies; volume is measured by multiplying
length by width by height

**a)**

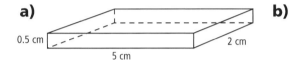

0.5 cm   2 cm
5 cm

**b)**

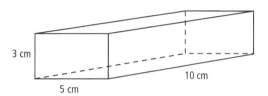

3 cm
5 cm   10 cm

**c)**

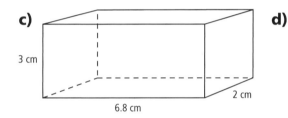

3 cm
6.8 cm   2 cm

**d)**

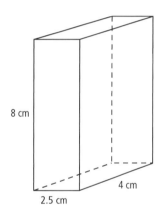

8 cm
2.5 cm   4 cm

**e)**

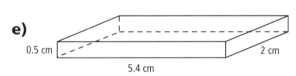

0.5 cm   2 cm
5.4 cm

**2.** There is a shorter formula for volume. In this version, the letter
*b* represents the area of the base, which is length multiplied by
width. For example, to find the volume for a rectangular prism
measuring 6 cm (length) by 4 cm (width) by 3 cm (height), you
would write the following:

$V = bh$

$V = 24$ cm $\times$ 3 cm

$V = 72$ cm$^3$

With a partner or in a small group, discuss this formula. Can
you develop another formula that would express volume
correctly? Work together to write your formula, and outline
how someone could apply it.

## What Did You Learn?

**1.** Which formula did you find easiest to use? Give reasons for your answer.

**2.** By now, you have had lots of practice with estimation. Imagine that a classmate was still having difficulty. List at least two pieces of advice you would give him or her to help improve their estimation skills.

## Practice

**1.** Record the formulas you used in this lesson.

**2.** Draw a rectangular prism. Label the length of its edges. Calculate its volume.

### Extension

**3.** Look at the figure below. Use the formula from Build Your Understanding question 2 to figure out the total volume of the figure.

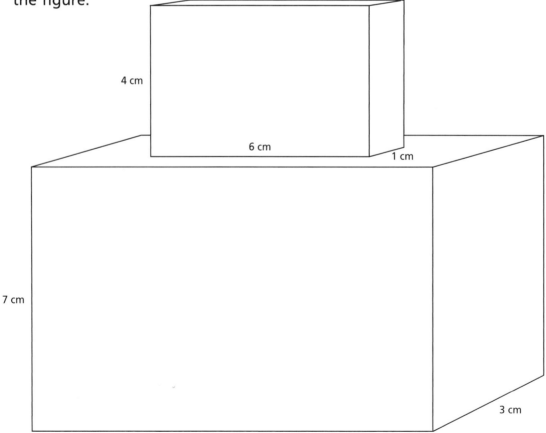

4 cm

6 cm

1 cm

7 cm

11 cm

3 cm

## Lesson 2

# Areas of Triangles and Parallelograms

*BUSINESS PLAN*

*PLAN:*
You will estimate and then find the areas of parallelograms and triangles.

*DESCRIPTION:*
With your flyer money, the first thing you decide to buy is a mini remote-control car. You want to discover the area of the car's triangular remote control as well as the box that the car and the remote came in.

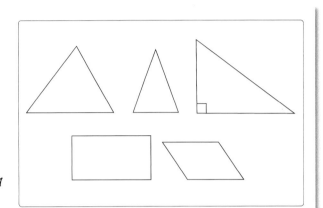

## Get Started

To find the area (*A*) of a triangle, multiply the base (*b*) of the triangle by its height (*h*) and then divide by 2. Here is the formula:

$A = b \times h \div 2$

Look at this example:

$A = b \times h \div 2$
$\quad = 6 \times 5 \div 2$
$\quad = 30 \div 2$
$\quad = 15$

The area (*A*) of the triangle is 15 cm$^2$.

RC Control

### Vocabulary

**area:** The amount of space inside a two-dimensional shape; area is measured in square units: mm$^2$, cm$^2$, dm$^2$, m$^2$, km$^2$
**perimeter:** The distance around a shape

The box your remote-control car came in is a parallelogram. To find the area (*A*) of a parallelogram, multiply the base (*b*) of the parallelogram by its height (*h*). Here is the formula:

$A = b \times h$

Here is one example:

$A = b \times h$

$\quad = 12 \times 6 = 72$

The area of the parallelogram is 72 cm².

Find the area of the triangle and the parallelogram shown on the right.

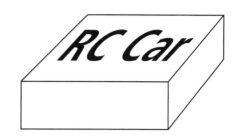

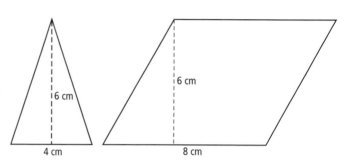

## Build Your Understanding

### Find Areas of Triangles and Parallelograms

1. Estimate and then use the formula $A = b \times h \div 2$ to calculate the areas of these triangles:

   **a)** $b = 16$ cm, $h = 4$ cm      **b)** $b = 3$ cm, $h = 4$ cm

   **c)** $b = 10$ cm, $h = 5$ cm      **d)** $b = 5$ cm, $h = 5$ cm

   **e)** $b = 1$ m, $h = 5$ m        **f)** $b = 3$ cm, $h = 8$ cm

2. Estimate and then use the formula $A = b \times h$ to calculate the area of these parallelograms.

   **a)** $b = 10$ cm, $h = 3$ cm      **b)** $b = 42$ cm, $h = 10$ cm

   **c)** $b = 75$ cm, $h = 70$ cm    **d)** $b = 32$ cm, $h = 43$ cm

   **e)** $b = 5$ cm, $h = 50$ cm     **f)** $b = 64$ cm, $h = 86$ cm

## What Did You Learn?

1. The formulas for finding the areas of parallelograms and triangles are similar but have one difference. Explain the reason for this difference.

2. Why are square units used for area?

**Journal**

Sketch possible shapes of triangles and parallelograms.

## Math Problems to Solve

Ajay is creating a pattern of equilateral triangles to make diamond shapes of four different colours. The base of each triangle is 5 cm. The area of the space he is covering is 1 m by 2 m.

1. How would he find the area he needs to cover?

2. How would he find the number of triangles he needs to cover the area? Record your work on paper.

3. When you are finished, trade papers with a classmate and compare your calculation methods.

4. Helene is wallpapering her room. The dimensions of the room are 3 m by 4 m by 3 m. How many square metres of wallpaper would Helene need to paper

   **a)** all four walls in her room?

   **b)** only the two longer walls?

   **c)** only the two shorter walls?

   **d)** one long wall?

**Booklink**

**Shape Up!: Fun With Triangles and Other Polygons** by David A. Adler (Holiday House: New York, NY, 1998). Various polygons including triangles, quadrilaterals, rhombuses, and dodecagons are introduced and illustrated by comparing them to common foods such as pretzels, wedges of cheese, and bread.

5. Draw a triangle. Now draw a rectangle that has the same base and the same height. Cut the triangle out and fit it into the rectangle. What can you say about the area of the rectangle in relation to the area of the triangle?

## Extension

6. Look at a variety of rectangles and parallelograms. Explain how you decide if a shape is a rectangle or a parallelogram.

## Lesson 3

# Measurement Relationships

BUSINESS PLAN

PLAN:
You will review different types of measurement—length, space, volume, time, and money—and how they can be related.

DESCRIPTION:
Measurement, in one form or another, is used in activities every day. You can probably think of a time when you had to use measuring skills while doing something ordinary or common in your daily schedule.

## Get Started

What types of measurement are shown in these illustrations?

**A**

**B**

**C**

CARPET
$9.99/m²

**D**

BAKERY
CLOSED

## Explore Measurement Concepts

**1.** Convert these units of measurement:

**a)** ▓ mm = 1 cm    **b)** ▓ cm = 1 m    **c)** ▓ m = 1 km

**d)** ▓ g = 1 kg    **e)** ▓ mm = 1 m    **f)** 2 L = ▓ mL

**g)** 3.5 L = ▓ mL    **h)** 3500 g = ▓ kg    **i)** 11 km = ▓ m

**2.** Volume is the product of length x height x width.
Here is the equation:

$V = l$ x $w$ x $h$

Use this equation to complete the following chart. Record your
answers in your notebook.

| Length | Width | Height | Volume |
| --- | --- | --- | --- |
| 18 cm | 7 cm | 3 cm | |
| 1 m | | 5 cm | 15 m³ |
| 15 mm | 12 cm | 3 cm | |
| 4 cm | 4 cm | | 64 cm³ |

**3.** Round these times to the nearest
half-hour.

**a)** 8:43 A.M.      **b)** 6:01 A.M.

**c)** 12:01 P.M.      **d)** 11:59 A.M.

**4.** Write these times as hours and minutes.
Round to the nearest minute.

**a)** 76 s      **b)** 365 min

**c)** 1001 min      **d)** 2 s

**Booklink**

**Counting on Frank** by Rod Clement (G. Stevens
Children's Books: Milwaukee, WI, 1991). A boy asks
questions about ordinary things including his dog,
Frank. He uses these questions to make math and
measurement fun and meaningful.

**5.** Round to the nearest dollar for each example.

a)

b)

c)

d)

## What Did You Learn?

As you have discovered in this lesson, you can represent one unit of measurement using another unit of measurement. Give an example of this for time and money.

## Math Problems to Solve

Li's brother has a job as a car jockey. He has a space of 45 m in which to park the cars and mini-bus shown below. For each problem, use pictures, numbers, and words to show your solutions. Compare the strategies you use with a partner's.

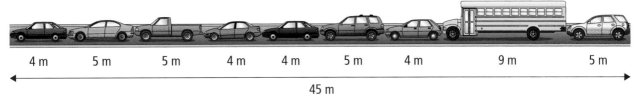

| 4 m | 5 m | 5 m | 4 m | 4 m | 5 m | 4 m | 9 m | 5 m |

45 m

1. Will he have enough space? Explain your thinking.

2. If he has to park the cars in two rows, what is the least amount of space he would need to park all of the cars and the mini-bus?

Look at question 2 of Build Your Understanding. Imagine that the numbers in the chart represent the measurements of several boxes in which you wish to pack some healthy snack bars. You have used some of your flyer money to buy them for a charity. Each snack bar is 15 cm in length, 2 cm in height, and 7 cm in width.

3. How many snack bars would you need to fill each of the boxes?

4. Are there any boxes that won't hold any bars at all?

Alana's grocery bill is $15.64. She pays the bill using 1 paper bill and 7 coins in her wallet. She receives 1 bill back.

5. What did Alana use to pay the grocery bill? Use numbers, pictures, and words to answer.

## Extension

6. With a partner, list ways in which each of these jobs might involve measurement:
   - firefighter
   - pharmacist
   - farmer

7. Identify another job and make a list of ways it might involve measurement.

## Lesson 4

# Surface Area and Volume

**BUSINESS PLAN**

**PLAN:**
You will find the surface area of a rectangular prism without using a formula. Then you will work with the volume of a prism and another solid object.

**DESCRIPTION:**
There is more than one way to find the area of a rectangular prism.

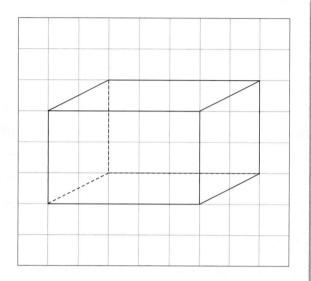

## Get Started

1. Draw a rectangular prism. Label all parts of the prism. Ensure that your diagram shows as many sides as possible.

2. Use your prism to make a design that could be used as the basis for a house. Use other geometric shapes in your work.

3. Display your work for others to view.

## Work With Surface Area and Volume

**You Will Need**
- rectangular prism
- 1-cm grid paper (optional)

**Vocabulary**

**surface area:** The sum of the areas of the faces of a three-dimensional object

### Calculate Surface Area of a Rectangular Prism

Imagine you are at a pet store to purchase a small aquarium with some of your flyer-delivery money. Fish need space to swim, so you need to be sure you don't buy too many fish for the size of aquarium you've bought. You'll need to measure the surface area of the aquarium to see how it will fit in your room. You will also need to estimate the aquarium's capacity so that you know how many fish you can buy. Use a rectangular prism to represent a smaller version of your aquarium.

1. Examine the prism. Look at its length, width, and height, and then estimate its surface area.

2. Trace the outline of each face of the prism on a piece of paper. Measure the length and width of each rectangle. (You can also use 1-cm grid paper for this step.)

3. Find the area of each rectangle.

4. Add the areas of all the rectangles to find the surface area of the prism. Record the surface area.

## Calculate Surface Area and Volume of a Full Aquarium

1. You decide to buy an aquarium that is 50 cm long by 20 cm wide by 30 cm high. Calculate the surface area of the aquarium to make sure it will fit the space you have.

2. Estimate the volume. Record your estimate.

3. Use the formula you learned ($V = l \times w \times h$) to calculate the volume of the prism.

4. Compare your estimate to the actual volume. If your estimate was not close to the volume, give reasons for the difference.

5. If each fish needs 3 L of water, how many fish can you buy?

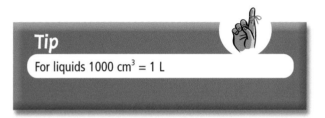

**Tip**

For liquids 1000 cm$^3$ = 1 L

**Vocabulary**

**capacity:** The amount of material an object, such as a container, can hold
**volume:** The amount of space that an object occupies; volume of a prism is measured by multiplying length by width by height

## What Did You Learn?

1. You have found the area of a rectangular prism in two ways. What is the value in finding area without a formula?

2. Explain how this lesson has helped you estimate volume.

## Practice

With a partner, review the concepts of volume that you have worked with in the past few lessons. List the procedures you use when you work with volume. Share your list with another pair of students.

## Lesson 5

# Mass and Volume

BUSINESS PLAN

PLAN:
You will estimate, measure, and record the mass and volume of rectangular prisms like the ones illustrated here.

DESCRIPTION:
Look at each set of boxes. Each larger box has double the volume of the smaller box. How can you tell?

## Get Started

**1.** Look at the prisms below. For each, draw a prism that would have twice the volume.

**a)**

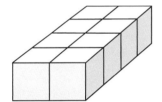

**b)**

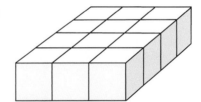

**2.** Estimate, then calculate the answers to these multiplication equations.

**a)** 9 x 8 x 1 x 2 = ▨

**b)** 6 x 5 x 3 x 4 = ▨

**c)** 9 x 2 x 8 x 2 = ▨

**d)** 4 x 2 x 3 x 3 = ▨

**e)** 4 x 3 x 4 x 3 = ▨

**f)** 5 x 5 x 5 x 3 = ▨

**g)** 50 x 10 x 3 x 3 = ▨

**h)** 12 x 6 x 3 x 2 = ▨

**i)** 25 x 3 x 5 x 5 = ▨

**j)** 2 x 3 x 4 x 5 = ▨

## Build Your Understanding

### Work With Mass and Volume

You Will Need
- rectangular prisms
- common scale
- 2 kg bag of sugar

You decide to help out with the family groceries by pitching in some of your flyer delivery money, and you want to figure out how much you're capable of carrying home with you. Your teacher will provide you with a number of rectangular prisms containing cubic centimetre blocks. These prisms represent the boxes of groceries. Work with a partner to complete the activities.

1. Hold the 2 kg bag of sugar in your hand to get a feel for its mass.

2. Choose three rectangular prisms. Based on the mass of the bag of sugar, estimate the mass of each prism, and record your estimate.

3. Check one prism's mass by placing it on a balance. Record its mass. How close was your estimate to the prism's mass? If your estimate was not accurate, think about reasons that could explain this.

4. Estimate the volume of the prism. What measurement unit will you use?

5. Calculate the volume of the prism and record it. Compare your estimate with the actual volume. Explain how you could make a more accurate estimate.

6. Repeat steps 2 to 4 for each of the three prisms.

**7.** Make a chart like the one below for the three prisms. For each prism, record these measurements: your estimate of mass, actual mass, your estimate of volume, length, width, height, and actual volume.

| Prism | Mass (Estimate) | Mass (Actual) | Volume (Estimate) | Length | Width | Height | Volume (Actual) |
|-------|-----------------|---------------|-------------------|--------|-------|--------|-----------------|
| 1 | | | | | | | |
| 2 | | | | | | | |
| 3 | | | | | | | |

**8.** How do changes in length, width, or height affect volume? Write a rule that expresses this relationship.

**9.** What does your chart show you about the relationship between mass and volume? With your partner, write a statement that summarizes your findings.

**10.** How would the mass change if the boxes were filled with marshmallows? Would this change the relationship between mass and volume?

## What Did You Learn?

Review the chart to compare your estimates against actual measurements. How accurate were your estimates? How can you improve your skill at estimating?

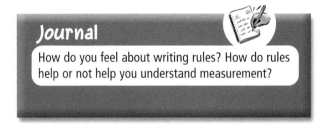

**Journal**

How do you feel about writing rules? How do rules help or not help you understand measurement?

# Practice

**1.** If you were to double the height of each of your three prisms, how would this affect the volume of each prism?

**2.** If you were to reduce the height of each of your three prisms by 50%, how would this affect the volume of each prism?

**3.** If you were to double the length of each prism, by what percentage would the volume increase?

## Extension

**4.** Make a model of a box that would contain all of the cubes that you were given in Build Your Understanding. Give your box to a classmate, and ask him or her to estimate and then calculate the volume of the cubes your box would hold.

**5.** Estimate the volume of your full desk. What measurement unit will you use? Why? With a partner, make a plan to find the volume of your desk. Then, calculate the volume of your desk and record it. Share and compare your answer with another pair of students. Why might your answers be different? How might you change your plan to get a more accurate answer?

### Show What You Know

Review: Lessons 1 to 5, Geometry and Measurement

**1.** Use linking cubes to create rectangular prisms of various volumes. Compare your prisms to a classmate's. How do they compare?

**2.** Draw three triangles on grid paper, each with a base of 4 cm and an area of 8 cm. Choose one triangle and change its base so that the area is 16 cm. Explain what you did.

**3.** How would you explain the difference between surface area and volume to someone who knew nothing about them?

**4.** Which masses are easier to lift?
a) 10 kg or 100 g
b) 1000 mg or 10 g
c) 1000 g or 1 kg

## Lesson 6

# Relationships of Terms

*BUSINESS PLAN*

*PLAN:*
You will find relationships between terms in sets of data and state these relationships as rules. You will use these rules to describe, complete, create, and extend patterns.

*DESCRIPTION:*
In the next few lessons, you will identify, extend, and create patterns. Learning about patterns can help you solve problems and make predictions in daily life.

## Get Started

1. You decide to buy some apples from a local fruit stand. You want to buy enough to last you two weeks so that you don't have to go to the fruit stand every day. Look at this illustration.

   What would you pay if you were to buy 4 apples? 5 apples? 6 apples?

2. Complete the following patterns:

   **a)** 2, 4, 6, ▨   **b)** 0, 3, 6, 9, ▨

   **c)** 2, 4, 8, ▨   **d)** 1, 2, 4, 5, 7, ▨

   **e)** 1, 3, 6, 10, ▨   **f)** 12, 9, 6, ▨

   **g)** 0, 3, 9, 18, ▨   **h)** 21, 32, 43, ▨

3. Complete the following patterns:

   **a)** A, b, C, ▨   **b)** a, c, e, g, ▨

   **c)** a, bb, ccc, ▨   **d)** c, a, d, b, ▨

   **e)** a, e, i, m, ▨   **f)** z, x, v, ▨

## Work With Data Tables

1. Work with a partner for this activity. Look at the sequences in question 2 of Get Started. For each sequence, discuss and identify the rule for the pattern. Do the same for the sequences in question 3 of Get Started. Describe each pattern rule.

2. Write four patterns consisting of at least 3 numbers or symbols. Make sure each pattern follows a different rule. Record your patterns in charts like the one shown below. Exchange your work with a classmate. Have him or her extend each pattern by four terms and record each rule. Check each other's work.

| Rule: | | | | | | | |
|---|---|---|---|---|---|---|---|
| | | | | | | | |

3. Extending the apple pattern from question 1 of Get Started, figure out how much 10 apples would cost. How much would 15 apples cost? On a graph, show the costs of 10, 11, 12, 13, 14, and 15 apples.

4. If you had $73.00 from your flyer-delivery job and all of your previous purchases totalled $60.00, how many apples could you buy?

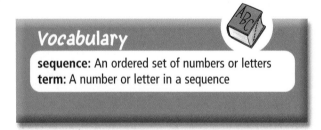

**Vocabulary**

**sequence:** An ordered set of numbers or letters
**term:** A number or letter in a sequence

1. Describe the strategies you used to complete the sequences in Build Your Understanding.

2. Describe how charts can help you identify patterns.

**1.** Write four patterns that each use a different operation (addition, subtraction, multiplication, and division). Exchange your work with a classmate. Have him or her extend each pattern and record each rule. Check each other's work.

**2.** With a partner, discuss the patterns you worked on in this lesson. Which patterns did you find easiest to solve? Which ones did you find most difficult to solve? Together, list strategies that could help you solve patterns in the future.

## Extension

**3.** Look for examples of patterns in magazines, newspapers, or flyers. Cut them out and glue them vertically on the left side of a piece of blank paper. Make sure you have permission to cut them out. Give your patterns to a classmate to extend. Display your patterns on a class bulletin board.

## Lesson 7

# Problem Solving With Patterns

BUSINESS PLAN

PLAN:
You will identify, extend, and create patterns dealing with money from your flyer-delivery business.

DESCRIPTION:
Patterns are not confined to numbers. They are a part of our everyday life—in the clothes we wear, the music we hear, and the graphics we see. How would you continue this pattern?

## Get Started

Identify the rule for the pattern below, then extend the pattern at least three terms. Record your answers in your notebook.

**1.** You open up a savings account at your bank. The first month, you deposit $10.00; one month later, you deposit $20.00; the next month, you deposit $30.00.

Develop a rule for the problem below. Sketch your answer in your notebook. Draw the pattern so that it repeats at least three times.

**2.** You decide to create a jacket for one of your CDs. As the jacket's background, you want to develop a print that is 25% blue, 25% pink, and 50% yellow. The pattern is an egg-shaped oval that repeats on a green background.

## Build Your Understanding

### Apply Patterning Strategies

There is a familiar children's song playing when you deposit your money at the bank. Math and music may seem very different, but they aren't. Each note is based on a number—the number of beats. To help you see the pattern, assign a letter or number to each note. For example, a quarter note could be A, a half note could be B. Write the song, substituting the letters or numbers for the notes. When you are finished, circle the repeating patterns in the song.

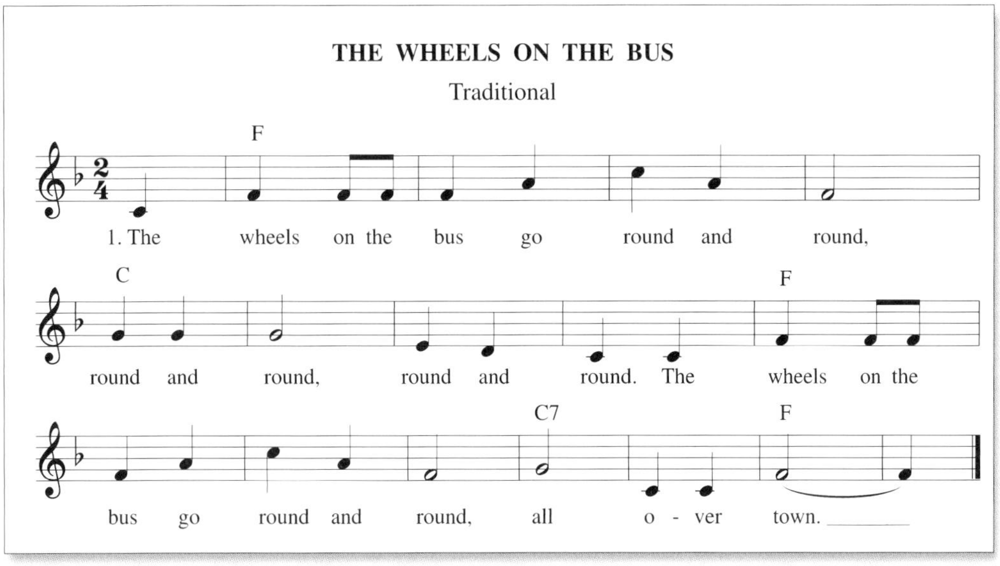


## What Did You Learn?

1. Use words or pictures to show the relationship between math and music.
2. Name a pattern that you see or use every day that you did not think of as a pattern before this lesson.

## Practice

Complete these patterns:

1. 0, 30, 15, 45, ■
2. 1, 20, 2, 40, 3, 60, ■
3. 2, 20, 4, 40, 8, ■
4. 1, 1, 2, 4, 3, 9, ■

# Math Problems to Solve

Solve these problems by showing them on a graph.

5. On your way to the bank, you notice patterns in the traffic passing you on a busy street. You estimate that you see an average of 12 cars per minute. Graph the number of cars you estimate you see in 5 min, 10 min, 15 min, 20 min, 25 min, and 30 min.

6. You find that your estimate in question 5 is 25% too high. Graph the number of cars you actually saw during these time periods.

> **Booklink**
>
> **Geometric Patterns from Roman Mosaics and How to Draw Them** by Robert Field (Tarquin Publications: Norfolk, England, 1999). Through photographs of Roman mosaics and grid drawings that show the development of pattern, the author provides a look at different art forms.

## Lesson 8

# Exploring Variables

BUSINESS PLAN

PLAN:

You will find the value of variables in simple formulas and continue your work with patterning.

DESCRIPTION:

At a local clothing store, you see a sign that says "4 shirts for $17.00." You want to buy only 3 shirts, so you need to divide the price of the 4 shirts by 4 in order to get the price of each shirt. Then you can multiply by 3 to get the amount you'll need to pay.

## Get Started

Janey works at a clothing store. The store is overstocked with shirts that need to be sold to make room for more stock. The store's owner offers his staff an extra $1.25 on top of their regular wages for every $20.00 worth of shirts they sell. During one 8-hour shift, Janey earns $26.25 extra. How many times did she sell $20.00 worth of shirts?

To find the number of times Janey sold more than $20.00 worth of shirts, you must solve for $n$ (the number of $20.00 sales). The guess-and-test method might help you. Start by writing your equation:

$26.25 = \$1.25 \times n$

Next, try $n = 20$

$1.25 \times 20 = \$25$ (not quite enough)

Try $n = 21$

$1.25 \times 21 = \$26.25$

Janey sold over $20 worth of shirts 21 times.

## Find Values of Variables

For the following problems, try the guess-and-test method first, and then check your work using a calculator.

Claire is one of Janey's customers at the clothing store.

1. The length of the rectangular box into which Claire puts her shirts is twice its width, plus 2 cm. The width is 14 cm. What is its length?

2. If she carries shirts home on her bike at a speed of 20 km/h, how far would she ride in 45 min?

3. Imagine you bought two boxes of tomatoes. The first box weighs half as much as the second, less 0.5 kg. The first box weighs 3 kg. How much does the other box weigh?

### Technology

With your teacher's approval, find programs on your school computer that you can use to construct patterns. Build as complex a pattern as you can. Print the pattern, and write the rule on the back of the page. Display your pattern with those of your classmates.

## What Did You Learn?

Tell a classmate how you found the missing term in question 1. Together, describe another way you could do it.

### Journal

Record what you found most surprising during your work with missing variables.

For each of these charts, fill in the missing terms. Write the equation for the rule. For example, if the rule is to add 1 to any given number ($x$), you would write this equation: $y = x + 1$. In other words, apply the rule to the number given for $x$. An example chart for this rule is

Rule: $y = x + 1$

| $x$ | 10 | 5 | 3 | 6 |
|-----|----|----|----|----|
| $y$ | 11 | 6 | 4 | 7 |

**1.** Rule:

| $x$ | 20 | 18 |    | 12 |    |    |
|-----|----|----|----|----|----|----|
| $y$ | 25 | 23 | 62 |    | 10 | 79 |

**2.** Rule:

| $x$ | 12 | 4 |    | 10 |    | 186 |
|-----|----|----|----|----|----|-----|
| $y$ | 36 | 12 | 60 |    | 93 |     |

**3.** Rule:

| $x$ | c | m |   | p |   | w |
|-----|----|----|----|----|----|----|
| $y$ | f | p | h |   | v |   |

## Show What You Know

### Review: Lessons 6 to 8, Patterning and Missing Terms

**1.** Create a pattern using numbers, then create a pattern using letters.

**2.** Explain the rules for both your number and letter patterns.

**3.** Which kind of pattern do you find more difficult to create? Why do you think that is?

**4.** Where might you or your family use patterning outside of school?

**5.** Complete these patterns:

   **a)** 4, 6, 5, 7, 6, 8, ▨

   **b)** b, e, h, k, ▨

   **c)** 10, 4, 9, 3, 8, 2, ▨

   **d)** a, z, b, y, c, ▨

Lesson 9

# Transformations

**BUSINESS PLAN**

*PLAN:*
**You will predict and represent transformations.**

*DESCRIPTION:*
In geometry, slides, flips, and turns are known as translations, reflections, and rotations. Each action is a type of transformation.

## Get Started

In geometry, a translation (or slide) means a figure moves up, down, or sideways on a page.

Look at this coordinate grid.

Notice the coordinates of the second figure. Notice, also, the prime symbols ('), which are added to show that a transformation has occurred.

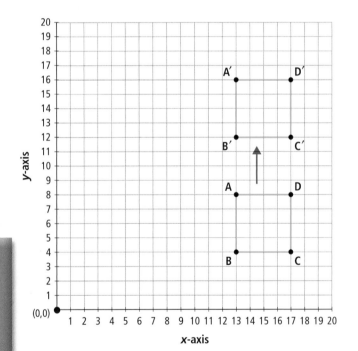

### Vocabulary

**reflection (flip):** A transformation (movement) of a figure by flipping it over a mirror line
**rotation (turn):** A transformation (movement) of a figure around a fixed point; the figure does not change size or shape, but it does change position and orientation.
**transformation:** A movement that does not change the size or shape of a figure
**translation (slide):** A transformation (movement) of a figure along a straight line

A reflection (or flip) means the figure has been flipped, as if taken from one hand and placed on the other, top side down.

Notice how the triangle has flipped just as an image in a mirror does.

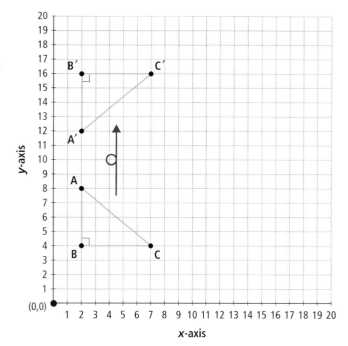

The final type of transformation is a rotation, or turn.

The rotation can be large (a full turn) or small (a quarter turn) either clockwise or counterclockwise.

**Tip**

**clockwise:** In the direction that the hands of a clock rotate
**counterclockwise:** In the direction opposite to the direction that the hands of a clock move

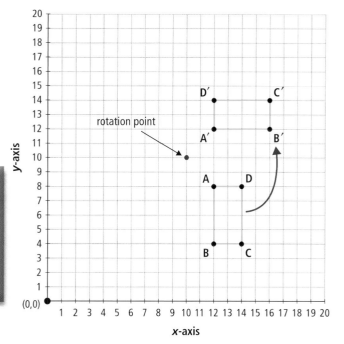

### Explore Translations, Reflections, and Rotations

You can trace these shapes onto a separate piece of paper and cut them out to help you visualize each transformation.

**1.** Give the coordinates for this figure if it were to slide to the right two squares and up five squares.

**2.** Give the coordinates for this figure if it were to make a half-turn around on point A in a clockwise direction.

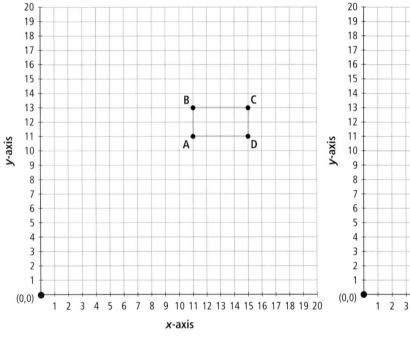

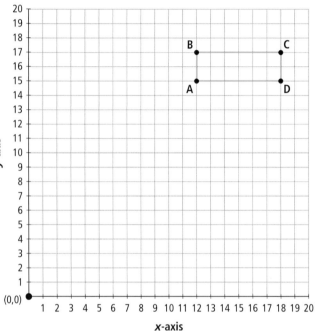

**3.** Give the coordinates for this figure if it were to flip to the right over the mirror line.

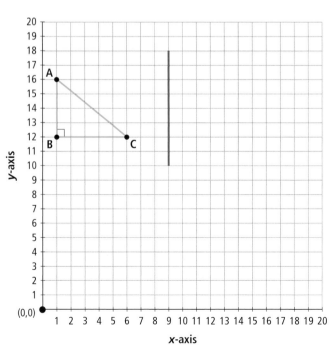

**4.** Identify the types of transformations that triangle ABC has gone through to become triangle ABC'.

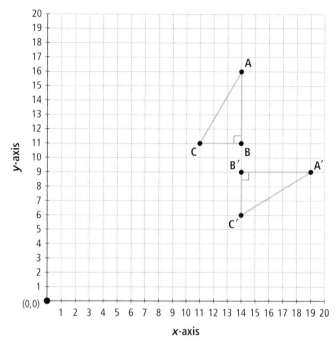

**5.** Identify the types of transformations that triangle ABC has gone through to become triangle ABC'.

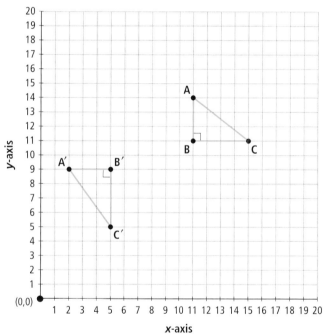

## What Did You Learn?

**1.** Without looking at your notes, define the terms translation, reflection, and rotation. Explain why they are described as transformations.

**2.** What changes about a shape when it is flipped? turned? slid?

**3.** Draw this figure on a coordinate grid. Make and label three more drawings to show three different transformations.

Lesson 9: Transformations    **415**

Look at each of the illustrations below. Identify the type of transformation(s) for each object.

**1.**

**2.**

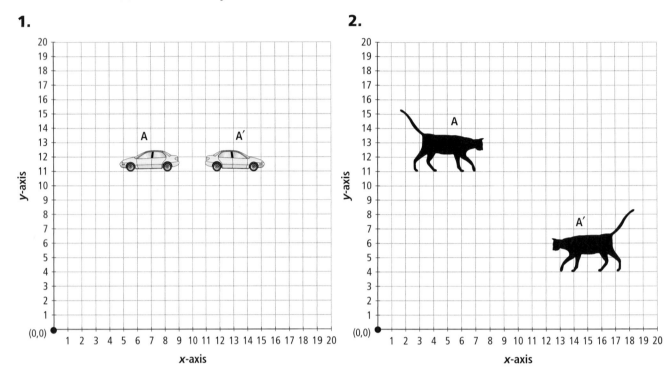

**3.**

**4.**

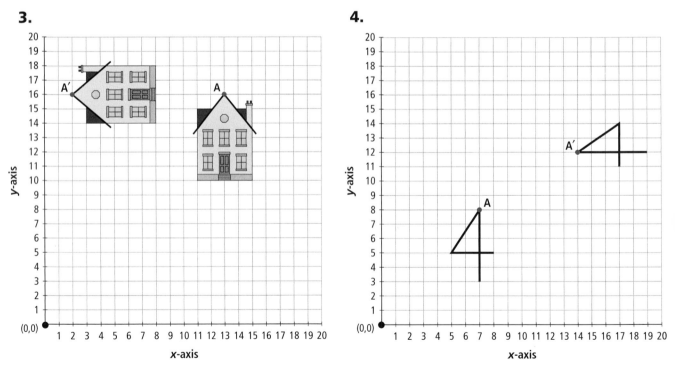

**5.** Describe two transformations that could move the original triangle to its new image position.

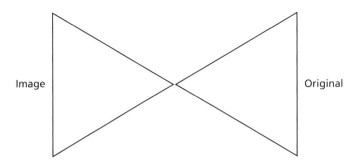

Image                    Original

Work with a partner. Using a book, perform the following transformations.

**6.** a translation (two book widths to the right)

**7.** a reflection (over the spine of the book)

**8.** a rotation of 90° clockwise

**9.** a rotation of 180° clockwise

**10.** a translation (one book width to the right) and a reflection (over the spine of the book)

**11.** a translation (two book widths to the left), a reflection (over the spine of the book), and a rotation of 45° counterclockwise

# Tessellations

## BUSINESS PLAN

PLAN:
**You will create a tessellation.**

DESCRIPTION:
Look at these illustrations. Each is an example of a tessellation (a repeating pattern of shapes). Some are called regular tessellations because they use only one shape.

Here are two examples of semi-regular tessellations. These types of tessellations use more than one shape. Which shapes make up each of these tessellations?

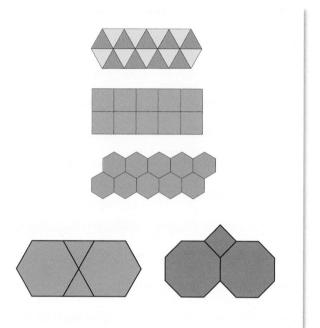

## Get Started

Look at these illustrations and statements. Identify each as a translation, a reflection, a rotation, or a combination. If it is a combination, describe it (for example, a rotation and a reflection).

1. A shape moves from the coordinates A (6, 4), B (6, 2), and C (3, 2) to A' (0, 4), B' (0, 2), and C' (3, 2).

**2.**

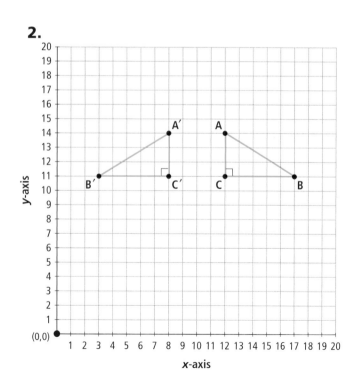

**3.** A shape moves from the coordinates A (3, 6), B (6, 6), C (6, 5), and D (3, 5) to A' (3, 4), B' (3, 1), C' (2, 1), and D' (2, 4).

Plot the points to draw the original shape and the image shape in order to help you.

## Build Your Understanding

## Explore Transformations of Geometric Figures

You Will Need
• a wooden geometric shape (optional)
• 1-cm grid paper

Decide which shape(s) you will use to make your tessellation, then use the guess-and-test method to create it.

**1.** Obtain a wooden geometric shape from your teacher, or draw a geometric shape on paper and cut it out.

**2.** Trace your shape on grid paper.

**3.** Perform a transformation on your shape.

**4.** Trace the new position of your shape. If you find that your transformation does not help you create a tessellation, continue trying different transformations until you find one that does help you create a tessellation.

**5.** Cover the entire paper with your tessellation. Record the type of tessellation you have created.

**6.** Once finished, record the types of transformations you used to make your tessellation.

### Vocabulary

**image:** The new position of a shape after a transformation. For example, for a point A transformed, the image point is shown as A' (read "A prime").
**regular tessellation:** A repeating pattern that uses only one shape, without gaps or overlaps
**semi-regular tessellation:** A repeating pattern that uses more than one shape, without gaps or overlaps

1. Think about rooms in your school or home. Where in these places can you find tessellations? Sketch and label some that are regular tessellations and some that are semi-regular tessellations.

2. In your own words, define regular tessellations and semi-regular tessellations.

**Practice**

1. Look at the tessellations other classmates created in Build Your Understanding. Group pieces according to whether they used a single transformation or more than one transformation. In each group, identify the transformation(s) used and the type of tessellation created.

2. From the work of other classmates, choose your favourite tessellation. Make a model of the shapes used and follow the description to make a smaller scale version of the tessellation.

### Extension

3. M.C. Escher, a Dutch graphic artist, is considered to be the master of tessellations. He used math to create works of art that still fascinate viewers today. Use the Internet to research his life and to view his works. When you are finished, answer this question: Was M.C. Escher an artist or a mathematician first? Once you have the answer, explain how you think this might have influenced his art.

**Booklink**

**Math Talk: Mathematical Ideas in Poems for Two Voices** by Theoni Pappas (Wide World Pub./Tetra: San Carlos, CA, 1991). This collection of 25 poems presents a variety of mathematical ideas including tessellations.

Chapter
**9**

# Chapter Review

**1.** Calculate the volume of the rectangular prism to the below. Show your work.

2.5 cm
4 cm
8 cm

**2.** Calculate the area of the shapes shown below. Be sure to show your work.

**a)**

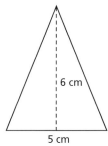

6 cm
5 cm

**b)**

3 cm

7 cm

**3.** Convert the following measurements:

**a)** ▨ mm = 15 cm      **b)** 4000 g = ▨ kg    **c)** 2 m = ▨ mm

**d)** 2 km = ▨ m         **e)** 3 L = ▨ mL       **f)** 360 min = ▨ h

**g)** 90 s = ▨ min

**4.** Calculate the volume of the rectangular prism to the right. Show your work.

**a)** What is the surface area of the prism?

**b)** If you double the height of the prism, what is the new volume? By what percentage did the volume increase?

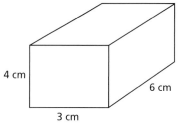

4 cm
6 cm
3 cm

**5.** Extend each of the following patterns by four terms:

**a)** 0, 4, 8, 12, ▨, ▨, ▨, ▨

**b)** 21, 18, 15, 12, ▨, ▨, ▨, ▨

**c)** 1, 2, 4, 7, 8, ▨, ▨, ▨, ▨

**d)** Explain the rule used in question c).

**6.** Draw a rectangle 2 cm wide and 8 cm long. Draw six more rectangles, each time increasing the width by 1 cm and decreasing the length by 1 cm. Record the area of each rectangle. Explain the pattern found in the area of these seven rectangles.

**7.** Complete these patterns:
   **a)** 0, 5, 3, 8, 6, 11, ■
   **b)** l, n, p, r, ■
   **c)** −0, 4, 8, ■
   **d)** 0, 2, 6, 12, 20, 22, ■
   **e)** Explain the rule used in question d).

**8.** Fill in the missing terms in the chart below. Write the equation for the rule.

| Rule: | | | | |
|---|---|---|---|---|
| *x* | 2 | 12 | 8 | |
| *y* | 24 | 144 | | 12 |

**9.** Using grid paper, draw a coordinate grid. Label and plot the following ordered pairs on your grid.
   **a)** (5, 3)   **b)** (6, 7)   **c)** (4, 6)
   **d)** (8, 2)   **e)** (7, 6)

**10.** Trace this equilateral triangle onto a coordinate grid. Illustrate a translation, a reflection, and a rotation Label each transformation.

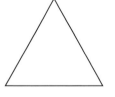

**11.** Using mathematical terminology, describe the transformations that make this pattern.

# Я Я R я Я

**12.** On grid paper, create a regular tessellation. Record the types of transformations your shape made to complete your tessellation.

# Chapter Wrap-Up

**Chapter 9**

In this chapter, you have covered a number of topics, including

- measurement concepts and their relationships
- area (parallelograms and triangles)
- surface area and volume (rectangular prisms)
- mass
- patterns
- coordinate grids
- transformations and tessellations

Now you can use the rules, formulas, and relationships you have worked with and studied. You will design CD jackets, then make and price containers for shipping them.

**1.** With your group, make a project plan.

- State your objective.
- Outline your project—what you will do and how you will do it.
- List the materials you will need.
- Identify a way to display your work.

**2.** Using the skills you learned in Chapter 9, make designs for your CD jackets. You could create tessellations, or use transformations to create different effects for various shapes, like parallelograms and triangles. Try to create fun and interesting patterns.

**3.** Once your CD jackets are created, you'll need to make packing containers for them. Decide within your group what materials should be used to pack your CD jackets. You'll then need to price your product. Estimate how many of your CD jackets will fit into your packing container, then use the formula for volume to calculate how much you'll have to charge for shipping.

**4.** Carry out your project by following your project plan. Keep notes or draw records of your work. When you have finished, prepare a visual presentation so that other groups can learn about your work.

**5.** In your notebook, describe the most positive and negative experiences you had while working as part of a group. Do you think the project would have been easier to complete on your own?

**6.** After viewing other groups' work, discuss with your group members what you have learned about the lessons in this chapter. Summarize how your group's work might help others.

**7.** From all of the groups' projects, prepare a summary or tip sheet for students starting Grade 6. This tip sheet should help them learn the math skills you have now learned. You can include rules, formulas, and illustrations to help you present your work.

# Problems to Solve

Here are some fun problems for you to solve. You get to choose the strategy you want to use for all of these problems.

**Problem 16**

## Relay Race Distance

STRATEGY: YOUR CHOICE

OBJECTIVE:
Calculate distance travelled

## Problem

Every Friday afternoon, the Grade 6 students participate in a relay race. The track is 500 m, and there are 4 students on each team.

1. How many teams would there be in your class?

2. How far will each student run if they each run the same distance?

Phillip, Ahanu, Li, and Ruth are on Team A.

3. How far would each student run in total if they ran this relay for three afternoons?

**Tip**

**Problem-Solving Steps**
1. Understand the problem
2. Pick a strategy
3. Solve the problem
4. Share and reflect

## Reflection

**1.** What did you know about the problem before you solved it?

**2.** What did you need to figure out?

**3.** What strategy did you use? Why?

## Extension

You Will Need
• calendar

If the four members of Team A ran this distance every Friday during April and May of this year, how far would they each have run in total?

**Problem 17**

# Prince Edward Island Population

*STRATEGY: YOUR CHOICE*

*OBJECTIVE:*

*Perform calculations involving whole numbers and percents*

## Problem

Imagine that you are doing a report on Prince Edward Island. For your report, you are trying to figure out the average number of people that live in 1 km². You found out that Prince Edward Island has an area of 5660 km². In 1999, Statistics Canada reported that 138 000 people lived on the island.

1. What was the population for 1 km² in 1999? You may use a calculator to help you.

2. If the population increased by 1% in 2001 and 2% in 2002, how many people would have been living on the whole island at the end of 2002? You may use a calculator to help you.

## Reflection

**1.** What did you know about the problem when you set out to solve it?

**2.** What did you need to figure out?

**3.** Which strategy did you use?

**4.** Do you think this strategy was a good choice? Why?

**5.** How will you check your answer?

## Extension

**1.** Research your own province, or a province of your choice, and figure out the population per 1 km$^2$.

**2.** Research the population of each of the Canadian provinces and arrange the populations in order from least to greatest.

### Technology

Visit the Statistics Canada Web site:
www.statcan.ca.

## Problem 18

# Fish Numbers

STRATEGY: YOUR CHOICE

OBJECTIVE:

Evaluate and interpret data. Express comparisons using ratios.

## Problem

Ahanu, Mahalia, Eric, and Ruth each have fish aquariums at home. Altogether, they have 20 tropical fish. 10 of the 20 tropical fish are white. Ahanu has 2 red and 1 yellow tropical fish. Ruth has 7 blue and 6 white fish. Mahalia only has white tropical fish. Eric has 1 white one and no blue or red ones.

1. Create a chart that compares by ratio the number of tropical fish each person has to the total number of fish. Write fractions and decimals for each ratio.

2. Write ratios to compare the number of white tropical fish that each person has to the total number of white fish. Write fractions and decimals for each ratio.

## Reflection

1. What did you need to figure out?

2. What did you know before you solved the problem?

3. What strategy did you use? Why?

4. How might you check your answers?

5. Share your answers with a classmate.

## Extension

Make up your own fish problem to share with a classmate.

**Problem 19**

# Blue Whale Mass

STRATEGY: YOUR CHOICE

OBJECTIVE:
Make calculations involving measurement

## Problem

Blue whales can grow to be over 33 m long, and they can eat up to 3600 kg of plankton per day. At this rate, how much plankton would a blue whale have eaten in 96 days? Use a calculator to help you.

## Reflection

1. Rewrite the problem in your own words.

2. What did you need to find out to solve the problem? What information was not necessary for you to solve the problem?

3. What strategy did you use? Why?

4. Why was this strategy a good choice?

5. Share your solution with a classmate.

## Extension

1. Make a problem that has to do with the masses of the following whales:

   Blue whale—95 000 kg        Humpback whale—32 000 kg

   Fin whale—48 000 kg         Grey whale—19 500 kg

   Sperm whale—36 000 kg

2. Research more math facts about whales and sea life. Share these facts with a classmate.

**Problem 20**

# Halley's Comet Pattern

*STRATEGY: YOUR CHOICE*

*OBJECTIVE:*
*Find a pattern*

## Problem

In 1705, English astronomer Edmond Halley predicted that a comet seen in 1682 would show itself again in 1758. When his prediction came true, the comet was named Halley's comet. Halley's comet was also seen in the years 1607, 1835, 1910, and 1986.

1. What pattern do you notice in the years that Halley's comet has appeared?

2. Predict which year Halley's comet will be seen next.

### Technology

Visit the NASA Web site: <www.nasa.gov>.

## Reflection

1. What did you know about the problem before you solved it?

2. What did you need to figure out?

3. Do you think your strategy was a good strategy to use? Why?

4. Discuss your solution with two classmates.

## Extension

Gather some math facts about space. Use the information you gather to make a math problem for a classmate to solve.

# Celebrating Math

Congratulations! You have almost finished the school year. You have learned a lot of new things this year, and now it's time to use what you have learned to celebrate. In Celebrating Math you will prepare for a class picnic. There are six fun activities that let you practise the math skills you have learned. Before you begin Celebrating Math, think about what math may be needed to plan a class picnic. Record your ideas and share them with your classmates in a small group.

At the beginning of the year, you were given some questions to answer. If you were asked those questions now, how would your answers be different from the first time? Share your responses with your classmates using all of the math knowledge you have gained this year.

**Lesson 1**

# Surveys and Calculations

PICNIC LOG

PLAN:
You will survey your class for their favourite picnic snacks and determine approximately how much money the food will cost.

DESCRIPTION:
A successful picnic should have a variety of foods to enjoy.

## Get Started

Design a survey to determine what kinds of snacks your classmates would like to have at the class picnic.

**1.** What snacks will you include on your survey?

**2.** How many of your classmates will you survey?

**3.** How will you organize your results?

## Build Your Understanding

### Calculate Costs

You Will Need
• number cube

The neighbourhood grocery store is having a sale. Snack foods are 25% off the regular price.

Work with a partner.

**1.** The numbers on the number cube correspond to the numbered snacks listed on the sign. Make a shopping

### Regular Prices

1. Pretzels: $1.99 a bag
2. Chips: $2.50 a bag
3. Freezies: $0.75 each
4. Ice-cream bars: $2.25 each
5. Pop: $1.25 a bottle
6. Cookies: $0.50 each

list by rolling a number cube six times to see which snack foods you will buy. For example, if you roll a 1 you will buy pretzels. Then, on your next roll, if you roll a 2 you will buy chips.

**Tip**

You may roll a number more than once, which means you will buy more than one of that particular item.

2. Calculate how much the regular price of each item is with 25% off.

3. How much money do you need to buy the groceries on your list?

4. How much change will you get if you pay with a $20.00 bill?

## What Did You Learn?

1. Arrange the foods from most expensive to least expensive according to the sale price.

2. Why is it important to know about percents?

3. According to your survey, which food was the most popular? Which food was the least popular?

## Practice

### Math Problems to Solve

1. The store is having a sale on ice-cream treats. The first ice-cream treat is on sale for $1.25. Each additional ice-cream treat bought costs $0.20 less than the previous one. How much will it cost to buy four ice-cream treats? Use numbers, pictures, and words to explain your answer.

2. Your teacher has brought a healthy snack of dried fruit to school. It consists of 30 raisins, 5 apricots, 35 bananas, and 15 dates. What is the ratio of dates to raisins? What other ratios can you show?

## Lesson 2

# Analyzing Data

PICNIC LOG

PLAN:
You will collect, organize, and analyze weather data.

DESCRIPTION:
The weather plays a big part in having a fun time at a picnic.

## Get Started

You Will Need
• grid paper

1. Look at the weather graph on the right.

2. Remake this weather graph in your notebook using different temperature intervals on the *y*-axis.

3. How is your scale different from the graph shown?

4. How has the appearance of your graph changed?

5. What season of the year do you think it is?

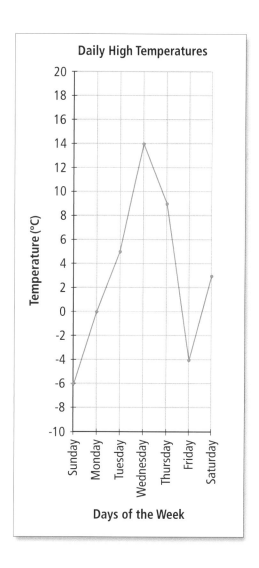

**Daily High Temperatures**

Temperature (°C)

Days of the Week

## Build Your Understanding

### Display and Analyze Data

**You Will Need**
- five-day weather forecast for your community
- grid paper

Work individually.

1. Display the high or the low temperatures for the five-day forecast two different ways.

2. Compare and contrast your two pieces of work.

3. Share your work with a classmate.

## What Did You Learn?

1. Look at your five-day forecast data and imagine that the picnic will be held on Friday. What is the probability of precipitation?

2. Based on the information you have, should the picnic go ahead as planned on Friday?

3. What is the median high temperature and the median low temperature for the five-day forecast?

4. Explain how you found the median.

5. What would be the best day for the picnic? Why?

## Practice

1. Make a stem-and-leaf plot to show these high temperatures in July:

| | | | | | | | |
|---|---|---|---|---|---|---|---|
| 14°C | 16°C | 17°C | 26°C | 7°C | 25°C | 21°C | 17°C |
| 25°C | 9°C | 21°C | 24°C | 16°C | 17°C | 25°C | 8°C |
| 26°C | 29°C | 25°C | 16°C | 21°C | 26°C | 26°C | 19°C |
| 28°C | 31°C | 19°C | 9°C | 24°C | 28°C | 13°C | |

2. Record the temperatures that are in the ranges of 0 to 9°C, 10 to 19°C, 20 to 29°C, 30°C and above.

3. In which range are the fewest temperatures found?

# Measurement Game

*PICNIC LOG*

*PLAN:*
You will create a game that involves linear measurement.

*DESCRIPTION:*
Picnics are wonderful times to play games.

## Get Started

Write a paragraph about what you have learned about the metric system of measurement. Be prepared to read your paragraph out loud to the class.

## Build Your Understanding

### Measure

You Will Need
- measuring tools (rulers, metre sticks, trundle wheels)
- balls
- rope
- large hoops

Work as a small group.

The picnic is fast approaching. Your task is to design a game to play at the picnic that uses the metric system of measurement. Record the rules of your game and how it is played. Share your ideas with your teacher, and get permission to try it out.

## What Did You Learn?

**1.** How is the metric system part of your game?

**2.** What would you change to improve your game?

## Practice

**1.** For the picnic, the students have a choice of wearing the following: a green, red, purple, or white shirt with blue jeans or black pants. What are all the possible combinations or outcomes of clothes the students could wear? Make a tree diagram to solve this problem.

**2.** Express five different measurements using: 3, 7, 1, 8. Each measurement must use all four digits, such as 1738 km.

Lesson 4

# Calculations

PICNIC LOG

PLAN:
*You will calculate the mathematical value of bars of music.*

DESCRIPTION:
Music is a fun way to liven up a picnic.

## Get Started

**1.** Copy these math sentences into your notebook. Then make each one true by filling in the blank.

**a)** 6 ■ 6 – 0 = 36 **b)** 5 x 5 – 7 + ■ = 28

**c)** 21 ÷ 3 + ■ = 22 **d)** ■ ÷ 6 – 5 = 3

**2.** Make up a few more questions like the ones above.

## Build Your Understanding

### Music Math

You Will Need
- number cube

Legend
Whole Note = ■ ■ ■ . ■

Half Note = ____________

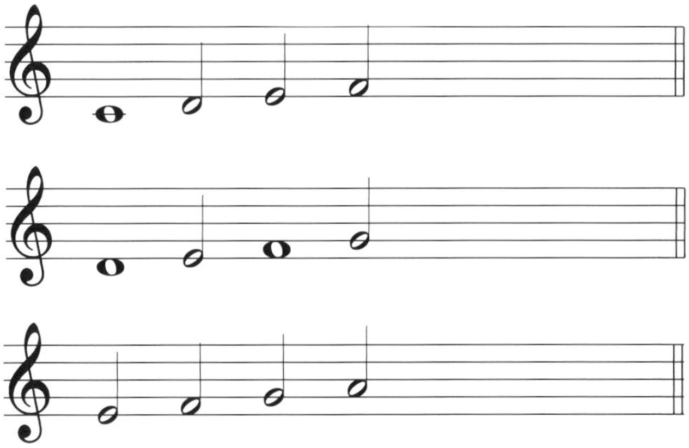

Work individually. Copy the notes and blanks above into your notebook.

**1.** Roll a number cube four times to fill in the blanks next to the whole note.

**2.** Calculate the value of the half note if it is half of the value of the whole note.

**3.** Use the numbers you have created to find the sum of the bars of music above.

**4.** Draw your own music bar with six notes, and calculate the sum.

## What Did You Learn?

**1.** Which bar of music had the greatest sum?

**2.** Which bar of music had the least sum?

**3.** How do you add numbers with decimals?

## Practice

**Square Strategy**

For example, 39 x 25

Start multiplying in the top right square.
9 x 2 = 18. Since 18 is a two-digit number, the tens digit goes in the left half of the square and the ones digit goes in the right half of the square.

Now multiply the top left square. Remember that the tens digit of the product goes in the left half of the square and the ones digit goes in the right. For one-digit numbers, there isn't a tens digit, so place a 0 in the left half of the square.

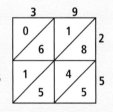

Add the numbers along the diagonal columns to find the answer, starting with the bottom right column. Sometimes the sum of the column will be a two-digit number. For example, in the question above, there are 17 tens. You know that 10 tens is 1 hundred, so you can trade 1 hundred to the hundreds column. You will have 7 tens left over in the tens column.

39 x 25 = 975

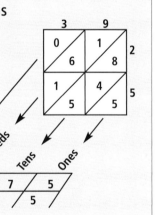

**1.** Use the "square strategy" to solve these multiplication questions. Begin by using your ruler to draw a square divided into quarters, and then into eighths with three diagonal lines, like the example shown above.

**a)** 42 x 16 = ■       **b)** 57 x 23 = ■

**c)** 69 x 31 = ■       **d)** 82 x 12 = ■

**2.** Make a few more questions like the ones above.

# Transformations

*PICNIC LOG*

*PLAN:*
*You will design a nature picture by using transformations.*

*DESCRIPTION:*
*Picnics are special because they give people a chance to be outside and enjoy nature.*

## Get Started

Explain and draw examples of a translation, reflection, and rotation.

## Build Your Understanding

### Transformations

You Will Need
- cardboard or thick paper
- scissors
- pens or markers

Work individually.

1. Choose a nature shape that is about the size of your hand, such as a leaf.

2. Draw your shape on cardboard and then cut it out.

3. Use your cardboard piece as a tracer and make transformations to create a picture.

4. Use pens or markers to colour your picture.

1. Look at your picture. What do you like about it? What would you change?

2. Describe the transformations you used to make your picture.

## Practice

Do the math to solve the puzzle.

| ■ | ■ | ■ | ■ | | ■ | ■ |
|---|---|---|---|---|---|---|
| 81 | 9 | 120 | 28 | | 42 | 5 |

| ■ | ■ | ■ | ■ | ■ | ■ | ■ | ■ | ■ | ■ |
|---|---|---|---|---|---|---|---|---|---|
| 8 | 6 | 8 | 180 | 132 | 49 | 28 | 8 | 180 | 8 |

9 x 9 = M                12 x 10 = T

6 x 7 = I                30 ÷ 5 = V

4 x 7 = H                10 x 18 = R

45 ÷ 9 = S               $8\overline{)64}$ = E

11 x 12 = Y              7 x 7 = W

$8\overline{)72}$ = A

# Class Picnic

You have reached the end of Celebrating Math. You have used your math skills to investigate and solve several math questions related to a class picnic. You learned that math is very important when planning a class picnic.

To celebrate the math you learned in this unit and throughout the year, imagine that you are a special-events coordinator who is responsible for making a detailed schedule of a class picnic. Your schedule should include math, such as the date and the start and end times of events. Be creative! There are many ways of incorporating math into your assignment. Here are a few questions to consider:

**1.** What food will be served?

**2.** When will people eat?

**3.** What games will be played?

**4.** What entertainment will be available?

**5.** Will there be a craft or art project?

**6.** When will the picnic start and end?

When you have finished your schedule, share it with a classmate. Go back and add two more ideas that came up while you were sharing.

# Glossary

**A.M.** Before noon; the time between midnight and noon

**acute angle** An angle that measures greater than 0° but less than 90°

Example:

**addend** Any number that is added

**area** The amount of space inside a two-dimensional shape; area is measured in square units: mm², cm², dm², m², km²

Example:

3 cm

3 cm

area = 9 cm²

**bar graph** A graph that uses parallel bars to show the relationship between quantities

Example:

**Favourite Colours**

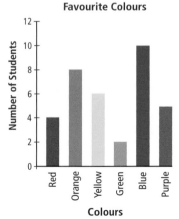

**base** A side of a polygon or a face of a solid figure by which the figure is measured or named; the symbol for base is *b*

Example:

base          base

**bias** An emphasis on characteristics not typical of the whole population; for example, if you were trying to find the most popular television program among Grade 6 students and only surveyed boys, your survey would be biased because it ignored girls' preferences

**bisect** Divided into equal halves

Example:

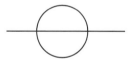

**C**

**capacity** The amount of material an object, such as a container, can hold

**centi** A prefix meaning one hundredth: $\frac{1}{100}$

**centimetre (cm)** $\frac{1}{100}$ of one metre

**centre of rotation** The point of intersection of two of the angle bisectors in a polygon or the point around which a polygon is turned

Example:

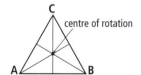

**circle (pie) graph** A graph that shows how parts of the data are related to the whole and to each other.

**clustering** A rule to change the value of like numbers to the closest common place value

**composite number** A whole number greater than 1 that has more than two factors; for example, 12 is a composite number because it has six factors

**congruent** Exactly the same size and shape; the symbol for congruence is ≡

Example:

**coordinate grid** A grid that has data points named as ordered pairs of numbers, such as (2, 3)

Example:

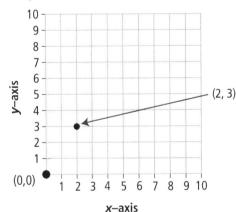

**coordinates** An ordered pair of numbers used to describe a location on a grid or plane; for example, the coordinates (2, 3) describe a location on a grid found by moving 2 units horizontally from the origin (0, 0) and then 3 units vertically

**cube** A three-dimensional solid with six congruent square faces

Example:

 **D**

**data** Facts or information such as statistics

**deci** A prefix meaning one tenth: $\frac{1}{10}$

**decimal** A number based on 10 with one or more digits to the right of the decimal point; for example, 0.36

**decimetre (dm)** $\frac{1}{10}$ of one metre

Example:

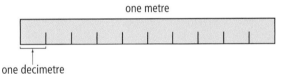

**denominator** The bottom part of a fraction that tells how many parts are in the whole; for example,

Example:

$$\frac{3}{4} \leftarrow \text{denominator}$$

**diagonal** A line segment that joins two non-adjacent vertices across a two-dimensional shape

Example:

**diameter** A line segment joining two points on the circumference of a circle and passing through the centre of the circle

Example:

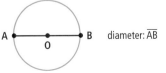

**difference** The answer to a subtraction question; for example, in the subtraction question 34 − 16 = 18, the difference is 18

**digits** Numerals from 0 to 9

**dimension** The length, width, or height of a figure

Example:

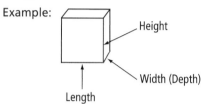

**discount** The amount by which the regular price is reduced to obtain the sale price; for example, a notebook might be $4.99. If it had a discount of $0.99, the sale price would be $4.99 − $0.99 = $4.00

**dividend** The number being divided; for example, in 56 ÷ 8, 56 is the dividend

**divisor** The number that divides the dividend; for example, in 56 ÷ 8, 8 is the divisor

**edge** The line segment where two faces of a three-dimensional figure meet

Example:

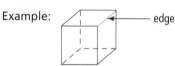

**equation** A mathematical statement where each side equals the other. The sides are separated by an equals sign; for example, 32 = 16 + 16

**equilateral triangle** A triangle with all sides equal in length

Example:

**equivalent fractions** Fractions that name the same amount or part; for example, $\frac{2}{4}$, $\frac{4}{8}$, and $\frac{3}{6}$ are all equivalent fractions; they can be reduced to $\frac{1}{2}$

**equivalent ratios** Ratios that represent the same fractional number or comparison; for example, 1:2, 2:4, 3:6

**expanded form** A number written to show the value of each digit; for example, 351 246 = 300 000 + 50 000 + 1000 + 200 + 40 + 6

**experimental probability** The chance of an event happening based on the results of an experiment

**face** One of the polygons of a solid figure

Example:

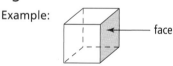

**factor** A whole number that is multiplied by another whole number to find a product; for example, 6 is a factor of 18 because 6 x 3 = 18

**formula** An open sentence that expresses a general rule; for example, the following are formulas for finding the perimeters of common polygons:

Rectangle: $P = 2l + 2w$

Square: $P = 4s$

Equilateral Triangle: $P = 3s$

Parallelogram: $P = 2l + 2w$

**fraction** A number that names a part of a whole; for example, $\frac{1}{2}$ is a part of one whole

**frequency table** A table that shows the number of times an event occurs

Example:

Frequency Table for Flipping
a Penny Experiment

| Event | Tally | Frequency |
|-------|-------|-----------|
| Heads | ⅲⅲ ‖ | 6 |
| Tails | ‖‖ | 4 |

**gram (g)** A metric unit used to measure mass

**greatest common factor (GCF)** The largest number that is a factor of two (or more) numbers; for example, 36 and 60 can both be divided evenly by 12. This is the GCF of both numbers.

**height** A line perpendicular to the base through the top vertex; the symbol for height is $h$

Example:

height

**hexagon** A polygon with six sides

Example:

**image** The new position of a shape after a transformation; for example, for a point A transformed, the image point is shown as A′

**improper fraction** A fraction with a numerator that is larger than the denominator

Example: $\frac{17}{4}$

**intersect** To have one or more points in common—intersecting lines cross each other at one point

Example:

**interval table** A table that shows the number of times an event occurs within an interval

**irregular polygon** A polygon with side and angle measurements that are not equal

Example:

**isometric dot paper** Dot paper whose dot pattern is formed by the vertices of equilateral triangles; it is used to make three-dimensional drawings

Example:
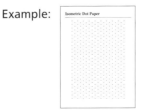

**isosceles triangle** A triangle that has two sides of equal length

Example:

3 cm    3 cm
2 cm

**K**

**kilo** A prefix meaning one thousand times (1000)

**kilogram (kg)** A metric unit, which is 1000 g

**kilometre (km)** 1000 metres

**kite** A quadrilateral with two pairs of equal adjacent sides

Example:

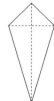

**L**

**least common multiple (LCM)** The smallest number, other than zero, that two or more numbers can divide into evenly; for example, the LCM of 2 and 3 is 6

**line graph** A graph that uses a line to show how data changes over a period of time

Example:

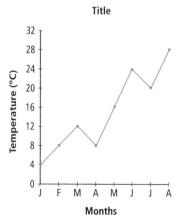

**line symmetry** Parts reflected in a line of symmetry

**line of symmetry** A line that divides a shape into two congruent symmetrical parts

Example:

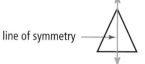
line of symmetry

**litre** A unit measure of capacity that is made up of 1000 mL

**lowest terms** A fraction or a ratio in which both the numerator and the denominator have no common factor other than 1; for example, $\frac{1}{3}$ is in lowest terms while $\frac{2}{6}$ is not, even though both fractions represent the same amount or part

**M**

**mass** The amount of matter in an object; usually measured in grams, milligrams, or kilograms

**mean** The average of a set of amounts. To calculate the mean, add up all the amounts and divide the sum by the total number of addends given. The mean of 2, 3, 5, 6, and 9 is 5.

**median** The middle value in a set of numbers arranged in order by size. The median of 1, 4, 7, 11, 15, and 20 is 10.

**metre (m)** A unit of measurement used to measure length

**milli** A prefix meaning one thousandth: $\frac{1}{1000}$

**milligram (mg)** A metric unit, which is $\frac{1}{1000}$ g

**millilitre** A unit measure of capacity; it is $\frac{1}{1000}$ of a litre and 1 mL = 1 cm³

**millimetre (mm)** $\frac{1}{1000}$ of one metre

**mixed number** A number that is part whole number and part proper fraction; for example, $7\frac{3}{4}$

**mode** The value that occurs most often in a set; the mode of 2, 3, 5, 5, 7, 5, 2, 8, 5 is 5

**multiple** The product of a given whole number and another whole number other than 1; for example, some multiples of 6 are 12, 18, and 24

**net** A two-dimensional pattern of a three-dimensional figure

Example:

**numerator** The top part of a fraction that tells how many parts are being referred to

Example:

$\frac{3}{4}$ ◄— numerator

**obtuse angle** An angle that measures greater than 90° but less than 180°

Example:

**octagon** A polygon with eight sides

Example:

**operation** A process in mathematics performed according to specific rules, such as addition, subtraction, multiplication, or division

**order of operations** The sequence in which complex equations are solved: multiplication and division first, then addition and subtraction, always working from left to right

Example:          2 + 3 x 4 – 5
                 = 2 + 12 – 5
                 = 14 – 5
                 = 9

**order of rotational symmetry** The number of times a tracing fits on the original figure during one 360° rotation around its centre of rotation

**ordered pair** A pair of numbers that form a location point on a coordinate grid; for example, (0, 2), (3, 4), (4, 5)

**origin** In a coordinate grid, the point where the *x*-axis and the *y*-axis intersect; it has the coordinates (0, 0)

**outcome** A possible result of a probability experiment

**P.M.** After noon; the time between noon and midnight

**parallel lines** Two lines that are continuously the same distance apart (they never cross)

Example:

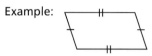

**parallelogram** A shape that has four sides with opposite sides that are parallel, and with opposite angles that are equal

Example:

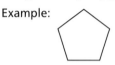

**pentagon** A polygon with five sides

Example:

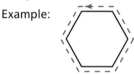

**percent** A number out of 100. It is written with the percent sign (%); for example, 18 percent is $\frac{18}{100}$ and is written as 18%

**perimeter** The distance around a shape

Example:

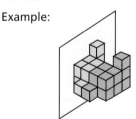

**plane** A flat or level surface that extends endlessly in all directions

**plane of symmetry** A plane that divides a solid into two congruent solids

Example:

**polygon** A shape that has at least three straight sides

Example:

**polyhedron** A three-dimensional figure with polygons for its faces

Example:

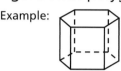

**population** The whole group of objects or individuals considered for a survey, such as all Grade 6 students in Canada

**possibility** All the outcomes that could occur; for example, when flipping a coin, two outcomes are possible: heads or tails; when rolling a number cube, six outcomes are possible: 1, 2, 3, 4, 5, or 6

**prime factor** A number that cannot be factored any further than 1 and itself. Examples of prime factors are 2, 3, 5, 7, 11.

**prime number** A whole number greater than 1 that has only two factors: 1 and itself; for example, 3 is a prime number because its only factors are 1 and itself

**prism** A three-dimensional figure with two faces that are congruent and parallel and other faces that are quadrilaterals

Example:

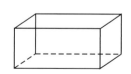

**probability** The chance or the likelihood that an event will happen. Probability can be expressed as a number from 0 to 1, where 0 means an even will certainly not occur and 1 means an event will definitely occur

**product** The answer to a multiplication question; for example, in the question 2 x 4 = 8, 8 is the product

**proper fraction** A fraction in which the numerator is smaller than the denominator; for example $\frac{3}{4}$

**pyramid** A solid figure with a polygon base and triangular faces that meet at a common vertex
Example:

**Q**

**quadrilateral** A polygon that has four sides
Example:

**quotient** The answer to a division question; for example, in the division question 36 ÷ 9 = 4, 4 is the quotient

**R**

**range** The difference between the greatest and least values in a group; for example, the youngest student in the school is 5, and the oldest student in the school is 13. 13 – 5 = 8. Therefore, the range is 8.

**rate** A ratio that compares two quantities with different units of measurement

**ratio** A comparison of two numbers with the same unit; for example, 1:4 means "1 compared with 4" and can be written as a fraction: $\frac{1}{4}$

**rectangular prism** A three-dimensional figure in which all six faces are rectangles; a square prism is a rectangular prism with two square faces
Example:

**reflection (flip)** A transformation (movement) of a figure by flipping it over a mirror line
Example:

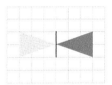

**reflex angle** An angle that measures between 180° and 360°
Example:

**regular polygon** A polygon with all sides and angles equal
Example:

**regular polyhedron** A polyhedron in which all the faces of the figure are congruent; for example, in a triangular-based pyramid, all faces are congruent equilateral triangles
Example:

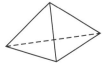

**regular tesselation** A repeating pattern that uses only one shape, without gaps or overlaps

Example:

**rhombus** A quadrilateral with four equal sides, and opposite angles that are equal

Example:

**right angle** An angle that measures exactly 90°

Example:

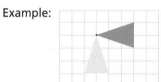

**rotation (turn)** A transformation (movement) of a figure around a fixed point (the figure does not change size or shape, but it does change position and orientation)

Example:

**rotational symmetry** When a shape can fit onto itself more than once during a full 360° turn

**rounding** A rule used to make a number approximate to a certain place value. Numbers are rounded up when the digit immediately following the required place value is 5 or higher, and rounded down when the digit is less than 5; for example, 238 rounded to the nearest 10 is 240; 1238 rounded to the nearest 100 is 1200

**sample** A part of a population selected to represent the population as a whole; for example, if the population is all Grade 6 students in Canada, a sample would be all Grade 6 students in your school

**sample space** The set of all possible outcomes

**scale** A device used for measuring mass

**scalene triangle** A triangle with three sides of different lengths

Example:

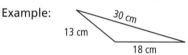

**scatterplot** A graph that uses plotted dots or points to show the relationship between two variables

Example:

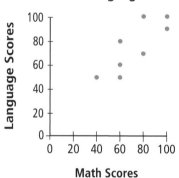

**semi-regular tessellation** A repeating pattern that uses more than one shape, without gaps or overlaps

Example:

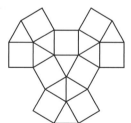

**sequence** An ordered set of numbers or letters

**SI** The International system of measurement units, which includes such metric units as millimetres and milligrams (SI is an abbreviation for the French terms "Systèm International d'Unités")

**similar** Shapes with the same angle measurements, but different, proportional side lengths; the symbol for similar is ~

Example:

**square dot paper** Dot paper whose pattern is formed by the vertices of squares

**square number** A whole number that is the product of another whole number multiplied by itself once. For example, 4 is a square number because it is the product of 2 x 2. 9 is a square number because 3 x 3 = 9.

Example:
```
•  •  •
•  •  •
•  •  •
```

**standard form** A number written as a numeral; for example, 351 246

**straight angle** An angle that measures exactly 180°

Example:

**sum** The answer to an addition question; for example, in the addition question 36 + 47 = 83, 83 is the sum

**surface area** The sum of the areas of the faces of a three-dimensional object

**survey** A method of gathering information by asking questions and recording people's answers

**symmetrical** Parts are balanced about a line or point

Example:

**tally chart** An organizer that uses tally marks to count data and record frequencies

Example:

| Fruit | Number of Students Who Like This Fruit |
|---|---|
| Apple | ⦀⦀⦀⦀⦀⦀⦀⦀⦀⦀⦀⦀ // |
| Banana | ⦀⦀⦀⦀⦀ // |
| Grapes | ⦀⦀⦀⦀⦀ //// |
| Pear | //// |

**term** A number or letter in a sequence

**terms of a ratio** The numbers used in a ratio; for example, the terms in the ratio 2:3 are 2 and 3

**tessellate** To cover an area with a repeating pattern, without gaps or overlaps

**theoretical probability** The number of favourable outcomes compared to the number of possible outcomes, when all outcomes are equally likely

**three-dimensional** Having three dimensions: length, width (depth), and height

Example:

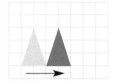

**transformation** A movement that does not change the size or shape of a figure

**translation (slide)** A transformation (movement) of a figure along a straight line

Example:

**trapezoid** A quadrilateral with only one pair of parallel sides

**tree diagram** A branching diagram that shows all possible outcomes of an event

Example:

```
                              Yellow
                   V-neck ────Green
                              Blue
T-shirts ──────── Buttons ──Striped
                              Purple
                   Collar ──── White
                              Beige
```

**triangular number** A whole number that can be represented by dots shown in a triangular array; for example, 1, 3, and 6 are the first three triangular numbers

Example:

**24-hour clock** A clock based on 24 hours, starting from 00:00 through 24:00

Example:

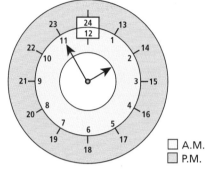

☐ A.M.
☐ P.M.

**two-dimensional** Having two dimensions: length and width but not height

Example:

width
length

# U

**unit rate** A ratio that compares a number to 1; for example, the delivery fee you earn for delivering one set of flyers

# V

**variable** A letter or symbol that is used in an equation to represent an unknown or missing value; for example 45 + $n$ = 51. In this equation, $n$ = 6

**vertex** The point where two rays of an angle or two edges meet in a plane figure, or the point where three or more sides meet in a solid figure (the plural of vertex is vertices)

Example:

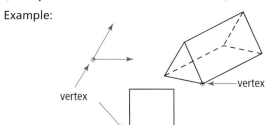

**volume** The amount of space that an object occupies; volume is measured by multiplying length by width by height

Example:

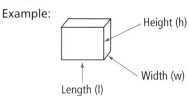

**whole number** The numbers belonging to the set [0, 1, 2, 3, 4 ....]

**x-axis** The axis that runs horizontally on a grid

**x-coordinate** The first number in an ordered pair; it tells the distance to move right or left from (0, 0). In (3, 2), 3 is the x-coordinate.

**y-axis** The axis that runs vertically on a grid

**y-coordinate** The second number in an ordered pair; it tells the distance to move up or down from (0, 0). In (3, 2), 2 is the y-coordinate.

# Problem-Solving Strategies

Here are some helpful strategies that you can use to solve the problems that appear in *Math Everywhere*.

Problem solving is a very important part of math. Problem solving allows you to practise the math skills you have learned.

Sometimes one strategy might not help you find the solution. If that happens, try another strategy. Trying different ways to find an answer is a part of learning.

1. Act out the problem to help visualize the solution.

2. Draw a picture to help you solve the problem.

3. Use objects like counters or play money to find the solution.

4. For large numbers, you can make a guess and then check to see if your guess is correct.

5. Begin with the information from the end of the problem and work backwards to find the solution.

6. For complicated problems, you can reduce large numbers to smaller numbers or reduce the number of items in the problem.

7. You can use concrete materials to build a model.

8. Looking for a pattern can help you solve many different kinds of problems.

9. You can make a table or a chart to help you organize information and find patterns in the data.

10. Making a list helps you organize information, too.

Here are some important steps to follow for all problems, no matter which strategy you use.

1. **Understand the problem:** Rewrite the problem in your own words. If you can, draw a picture of the problem. List or highlight important numbers or words.

2. **Pick a strategy:** You will be learning about many different problem-solving strategies throughout the year. For example, "Act It Out," "Draw a Picture," "Use Objects," and "Guess and Check."

3. **Solve the problem:** Use a strategy to solve the problem. Describe all steps using math words and/or symbols. Try a different strategy if you need to. Organize the results using a picture, model, chart, table, or graph.

4. **Share and reflect:** Did the strategy you picked work? Would a different strategy also work? Does your solution make sense? Could there be more than one answer to the problem? How did other people in your class solve the problem?

## Acknowledgements

*Photographs*

T = Top, C = Centre, B = Bottom, L = Left, R = Right, M = Middle

p. 20: Mehau Kuluk/Science Photo Library/Publiphoto; 21: NASA/Science Photo Library/Publiphoto; 22: Science Photo Library/Publiphoto; 25: (l) NASA/Science Photo Library/Publiphoto, (r) © Tom Van Sant/Geosphere Project, Santa Monica/Science Photo Library/Publiphoto; 29: Digital Vision/Getty Images; 31: Jim Schwabel/MaXx Images; 37: NASA/Science Photo Library/ Publiphoto; 40: NASA/SPL/Publiphoto; 43: European Space Agency/Science Photo Library/ Publiphoto; 47: NASA; 48 NASA/Science Photo Library/Publiphoto; 54: NASA/Science Photo Library/Publiphoto; 57: NASA/Science Photo Library/Publiphoto; 60: CP(Alan Diaz); 68: Ron Kocsis/Publiphoto; 72: © George Shelley/CORBIS/MAGMA; 75: NASA/Science Photo Library/ Publiphoto; 79: John Sanford/Science Photo Library/Publiphoto; 81: F. Gohier/Photo Researchers/Publiphoto; 89: Chris Bjornberg/Photo Reserachers/Publiphoto; 96: NASA; 99: JPL/NASA; 106: Space Telescope Science Institute/NASA/Science Photo Library/Publiphoto; 107: (tl) NASA/Science Photo Library/Publiphoto, (tr) Tom Van Sant/Geosphere Project, Santa Monica/Science Photo Library/Publiphoto, (bl) U.S. Geological Survey/Science Photo Library/ Publiphoto, (br) Chris Bjornberg/Photo Researchers/Publiphoto; 108: (t) U.S. Geological Survey/ Science Photo Library/Publiphoto, (m) NASA/Science Photo Library/Publiphoto, (m) Tom Van Sant/Geosphere Project, Santa Monica/Science Photo Library/Publiphoto, (b) Chris Bjornberg/ Photo Reserachers/Publiphoto; 110: © Gunter Marx Photography/CORBIS/MAGMA; 113: Chris Butler/Science Photo Library/Publiphoto; 115: © Superstock; 117 NASA; 120: Science Photo Library/Publiphoto; 123: NASA; 124 (t) NASA/Science Photo Library/Publiphoto, (b) Image Bank/Getty Images; 127: NASA; 131: (t) A. Tayfun Oner, (tl) U.S. Geological Survey/Science Photo Library/Publiphoto, (ml) NASA/Science Photo Library/Publiphoto, (m) © Tom Van Sant/Geosphere Project, Santa Monica/Science Photo Library/Publiphoto, (m) © Chris Bjornberg/Photo Reserachers/Publiphoto, (b) Science Photo Library/Publiphoto, (tr) NASA/ Science Photo Library/Publiphoto, (mr) NASA, (mr) Southern Stock/Maxx Images, (br) Image Bank/Getty Images; 134: Lynette Cook/Science Photo Library/Publiphoto; 140: Science Photo Library/Publiphoto; 148: © Oliver J. Troisfontaines/SuperStock; 150: Courtesy of Canada Post; 154: Bill Ivy/Ivy Images; 155: Digital Vision/Getty Images; 159: © Royalty-free/CORBIS/MAGMA; 163: © Jack Fields/CORBIS/MAGMA; 167: (t) Bill Ivy/Ivy Images, (b) © Royalty-free/CORBIS/ MAGMA; 170: Bill Ivy/Ivy Images; 174: Courtesy of Air Canada; 179: Tim Pelling/Ivy Images; 186: © Ruet Stephane/CORBIS SYGMA/MAGMA; 195: © David Lawrence/CORBIS/MAGMA; 196: © Charles O'Rear/CORBIS/MAGMA; 202: Courtesy Calgary Airport Authority; 206: Tim Pelling/Ivy Images; 207: Photodisc/Getty Images; 211: Courtesy Calgary Airport Authority; 215: W. Fraser/Ivy Images; 219: Courtesy Calgary Airport Authority; 229: (t) © Benjamin Rondel/ CORBIS/MAGMA, (b) © George Hall/CORBIS/MAGMA; 234: CP/Toronto Star (Tony Bock); 238: (t) Ivy Images, (b) W. Lowry/Ivy Images; 247: © Superstock; 249: (tl) © Underwood & Underwood/CORBIS/MAGMA, (bl) © Royalty-free/CORBIS/MAGMA, (r) © CORBIS/MAGMA; 250: Bill Ivy/Ivy Images; 255: © Roger Allyn Lee/Superstock; 266: © Baldwin H. Ward & Kathryn C. Ward/CORBIS/MAGMA; 270: Don Stevenson/MaXx Images; 274: © Robert Llewellyn/Superstock; 278: (l) Bill Ivy/Ivy images, (m) © ThinkStock/SuperStock, (r) Dick Hemingway; 279: Myrleen Cate/MaXx Images; 283: © Bettmann/CORBIS/MAGMA; 288: Canadian Aviation Museum; 296: © Bill Ivy/Ivy Images; 301: (t) Ottmar Bierwagen/Ivy Images, (b) Ottmar Bierwagen/Ivy Images; 303: MaXx Images; 304: © Kennan Ward/CORBIS/MAGMA; 306: Bill Ivy/Ivy Images; 427: Phyllis Picardi/MaXx Images; 429: Shedd/Cecil/Visuals Unlimited; 430: Flip Nickin/Minden/firstlight.ca; 431: Victoria Johana/MaXx Images

*Illustrations*

Allan Clarke: pp. 407, 439; Deborah Crowle: pp. 99, 237, 261 (t), 288, 312 (t), 348, 350 (t), 394, 395, 416; Susan Gardos: pp. 1, 2, 12, 13, 15, 16, 17, 26, 28, 33, 37, 41, 44, 55, 62, 67, 82 (b), 96, 105, 110, 120, 125, 130, 133, 137 (t), 145, 151, 153, 161, 182, 186, 188, 197, 212, 222, 225, 265, 270, 297, 300, 307, 308, 311, 316, 322, 326, 328, 329, 334, 336, 346, 347, 352, 354, 355, 368, 371, 372 (b), 374, 375 (b), 385, 386, 391, 392 (t), 397, 400, 403, 408, 409, 410, 412, 423, 425, 432, 433 (t), 437, 441, 443; Stephen Hutchings: pp. 18, 51, 53, 58, 82 (t), 86, 147, 223, 232, 285, 298, 312 (b), 319, 332, 339, 344, 350 (b), 362, 369, 372 (t), 376, 379 (b), 392 (b), 406; Jock MacRae: pp. 8, 10, 38, 69, 137 (b), 156, 166, 280, 283, 389; Liz Milkau: pp. 168, 202, 221; Dorothy Siemens: pp. 6, 72, 75, 80, 85, 93, 309, 313, 317, 318, 365, 375 (t),  378, 379 (t), 399, 433 (b).

Technical Illustrations: Deborah Crowle, Jock MacRae
Icons: Carl Weins

The authors and publisher gratefully acknowledge the contributions of the following educators in the development of *Math Everywhere:*

**Michael Beetham**
TEACHER, Westmount Public School
Waterloo District School Board
Kitchener, Ontario

**June Buick**
VICE PRINCIPAL, Our Lady of Fatima School
York Catholic District School Board
Woodbridge, Ontario

**Rita Cardarelli**
ELEMENTARY CURRICULUM RESOURCE TEACHER
Ottawa-Carleton District School Board
Napean, Ontario

**Josephine Carnevale**
TEACHER LIBRARIAN, St. Edith Stein School
Dufferin-Peel Catholic District School Board
Mississauga, Ontario

**Joseph DiFrancesco**
PRINCIPAL, Our Lady of Fatima
Brant Haldimand Norfolk Catholic District School Board
Brantford, Ontario

**Dana Free**
TEACHER, H. W. Knight Public School
Durham District School Board
Cannington, Ontario

**Wendy Gallant**
VICE PRINCIPAL/MATH SUBJECT LEADER
Algonquin and Lakeshore Catholic District School Board
Kingston, Ontario

**Wes Hahn**
SPECIAL ASSIGNMENT TEACHER
Hamilton-Wentworth District School Board
Hamilton, Ontario

**Colleen MacDonald**
CONSULTANT
Ottawa-Carleton Catholic District School Board
Napean, Ontario

**Judy Mendicino**
TEACHER, Manchester Public School
Waterloo District School Board
Cambridge, Ontario

**Debbie Schwantz**
TEACHER, Woodland Park Public School
Waterloo District School Board
Cambridge, Ontario

**Theresa Spencer**
SPECIAL ASSIGNMENT TEACHER
Sudbury Catholic District School Board
Sudbury, Ontario

**Dianne Phillips**
CONSULTANT
Upper Canada District School Board
Prescott, Ontario

**Mary Beth Yahn**
TEACHER, Algonquin Ridge Elementary School
Simcoe District School Board
Midhurst, Ontario

**Stephanie Bishop**
TEACHER, Geary Elementary School
School District 17
Oromocto, New Brunswick

**Susan Brims**
TEACHER, West Dalhousie Elementary School
Calgary Public Schools
Calgary, Alberta

**Bonnie Chappell**
DIRECTOR, CURRICULUM AND INSTRUCTION
School District 57
Prince George, British Columbia

**Cindy Coffin**
MATH & LANGUAGE ARTS CONSULTANT
Saskatoon Catholic Schools
Saskatoon, Saskatchewan

**Ruth LeBlanc**
MATH MENTOR
Moncton School District 2
Riverview, New Brunswick

**Denise McWilliams**
TEACHER/CONSULTANT
River East School Division
Winnipeg, Manitoba

**Darren McMillan**
CONSULTANT
School District 43
Coquitlam, British Columbia

**Suzanne Prefontaine**
SPECIALIST MATH TEACHER,
Holyrood Elementary
Edmonton Public Schools
Edmonton, Alberta

**John Price**
CURRICULUM CONSULTANT
School District 39
Vancouver, British Columbia

**John Pusic**
COORDINATOR—INSTRUCTIONAL SERVICES
School District 35
Langley, British Columbia

**Deb Scott**
TEACHER, Hastings School
Louis Riel School District
Winnipeg, Manitoba

**Joanne Stubbs**
ELEMENTARY MATH AND SCIENCE CONSULTANT
Prince Edward Island Department of Education
Charlottetown, Prince Edward Island

**Marilyn Wolstenholme**
TEACHER, Florence Elementary School
Cape Breton-Victoria School Board
Florence, Nova Scotia

**Tammy Wu**
TEACHER, Caulfield School
School District 45
West Vancouver, British Columbia